T0351278

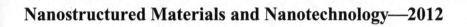

Nanostructured Materials and Nanotechnology—2012

MATERIALS RESEARCH SOCIETY
SYMPOSIUM PROCEEDINGS VOLUME 1479

Nanostructured Materials and Nanotechnology—2012

Symposium held August 12–17, 2012, Cancún, México

EDITORS

Claudia Gutiérrez-Wing

Instituto Nacional de Investigaciones Nucleares
Ocoyoacac, México

José Luis Rodríguez-López

Instituto Potosino de Investigación Científica y Tecnológica
San Luis Potosí, México

Olivia A. Graeve

Department of Mechanical and Aerospace Engineering
University of California, San Diego, USA

Milton Muñoz-Navia

Universidad de La Cienega del Estado de Michoacán de Ocampo
Michoacán, México

Materials Research Society
Warrendale, Pennsylvania

CAMBRIDGE
UNIVERSITY PRESS

CAMBRIDGE
UNIVERSITY PRESS

University Printing House, Cambridge CB2 8BS, United Kingdom

One Liberty Plaza, 20th Floor, New York, NY 10006, USA

477 Williamstown Road, Port Melbourne, VIC 3207, Australia

314-321, 3rd Floor, Plot 3, Splendor Forum, Jasola District Centre, New Delhi - 110025, India

103 Penang Road, #05-06/07, Visioncrest Commercial, Singapore 238467

Cambridge University Press is part of the University of Cambridge.

It furthers the University's mission by disseminating knowledge in the pursuit of education, learning and research at the highest international levels of excellence.

www.cambridge.org
Information on this title: www.cambridge.org/9781605114569

© Materials Research Society 2013

First published 2013

CODEN: MRSPDH

A catalogue record for this publication is available from the British Library

ISBN 978-1-605-11456-9 Hardback

CONTENTS

PREFACE

The emerging field of Nanotechnology has envisioned a great impact in basic and applied research in different areas, leading to the production of materials with novel properties and devices with new or improved performance. Studies on nanostructured materials and phenomena occurring at the nanoscale are fundamental elements for the development of nanotechnology. Findings in this field have opened new opportunities for active research in different areas such as biomedicine, catalysis, electronics, and cosmetics, among others.

With the continuous aim of providing a forum where the latest developments in nanoscience and nanotechnology can be presented and discussed, the XXI International Materials Research Congress held in Cancún, México in August 2012, hosted the Nanostructured Materials and Nanotechnology Symposium. As in previous editions of the congress, this symposium has served as a stage for the presentation of oral and poster contributions in the field, were we can learn of the advancement of Nanoscience and Nanotechnology considering experimental and theoretical approaches. Participants from different countries and a great amount of contributions from México have taken part of the development of this congress, covering a number of topics which include the synthesis of nanostructures and nanocomposites; optical, electrical and structural characterization techniques; modeling of structures and theoretical analysis of properties; carbon based nanostructures and different applications of nanomaterials as in catalysis, biomedicine and sensors development.

In this volume, we have compiled a number of interesting contributions which represent some of the many topics presented in the symposium. We thank the authors of the papers presented here and the important support of those who made possible the preparation of this volume.

Sincerely,

The Editors

MRS Proceedings / Nanostructured
Materials and Nanotechnology Symposium - XXI IMRC

MATERIALS RESEARCH SOCIETY SYMPOSIUM PROCEEDINGS

MATERIALS RESEARCH SOCIETY SYMPOSIUM PROCEEDINGS

MATERIALS RESEARCH SOCIETY SYMPOSIUM PROCEEDINGS

Volume 1534E — Low-Dimensional Semiconductor Structures, 2012, T. Torchyn, Y. Vorobie, Z. Horvath,
ISBN 978-1-60511-511-5

Prior Materials Research Society Symposium Proceedings available by contacting Materials Research Society

Mater. Res. Soc. Symp. Proc. Vol. 1479 © 2012 Materials Research Society
DOI: 10.1557/opl.2012.1589

Effect of Equivalent Sites on the Dynamics of Bimetallic Nanoparticles

C. Fernández-Navarro[1], A. J. Gutiérrez-Esparza[2], J.M. Montejano-Carrizales[3] and S.J. Mejía-Rosales[1]

[1]Facultad de Ciencias Físico-Matemáticas, Universidad Autónoma de Nuevo León, San Nicolás de los Garza, Nuevo León, México 66450
[2]División de Ciencias e Ingenierías, Universidad de Guanajuato, León, Gto. México 37150
[3]Instituto de Física, Universidad Autónoma de San Luis Potosí, San Luis Potosí, S.L.P., México 78000

ABSTRACT

Using a Sutton and Chen interatomic potential, we study the molecular dynamics of Au-Pd nanoparticles with an initial icosahedral structure at different temperatures and concentrations, where each relative concentration of the 561-atom particles was made by placing atoms of the same species at equivalent sites, in order to identify under which conditions the melting transition temperature appears for each particle. In addition, we compute global order parameters in order to correlate the obtained results with the caloric curves of each particle. As a result, we observe that the melting transition temperature depends on the relative atomic positions of gold and palladium. The melting transition temperature of the Au-Pd alloy particles appears at higher temperature than that of the pure-gold particle. From the analysis of the structure of the particles, we found that the melting temperature increases with the proportion of gold atoms, and for those particles with a higher concentration of palladium on the surface, we observe an early migration of gold atoms before the melting transition temperature appears.

Keywords: Nanoparticles, gold, palladium, molecular dynamics.

INTRODUCTION

The uses of nanoparticles have grown in a large measure both in number and scope over the last decade. Besides their use as catalysts, metallic nanoparticles are very promising agents in the development of electronic devices. Nowadays the application of nanostructured materials as protein detectors [1] contrast enhancers in imaging techniques [1], catalysts [2], drug delivery [3], and antibacterial agents [4], has become of great importance, mainly due to their stability and selectivity features. As applications of nanoparticles grow in different and sometimes unexpected directions, the need of a better understanding of the influence of structure on the properties of the particles becomes more accentuated, specially when the choice of a method of synthesis is determined by how the final structure and size of the nanoparticle depend on the chosen technique [5-7]. Synthesis is possible with chemical based methods to obtain colloidal dispersions of nanoparticles [8] and physical methods of deposition, where the nanoparticles are formed by high-pressure sputtering in a controlled gas flow atmosphere, can be used to generate particles with narrow size distributions and, more important, with well defined chemical composition [9,10].

Along with the variety in the methods of synthesis, characterization techniques have been improving constantly. One of the most important tools in the analysis of nanostructures is high-resolution transmission electron microscopy (HRTEM) [11]; with HRTEM is possible to study simultaneously the structure and the chemical composition of the particles with a high degree of resolution. By the use of numerical methods is possible to make the reconstruction of electron exit waves from focal series of lattice images, what extends the HRTEM resolution to fractions of Angstrom, allowing imaging the atomic structure of the nanoparticle directly. This resolution capability potentiates the in-depth study of the geometrical characteristics of the nanoparticles, and their differences with respect to the crystal structure of the bulk.

Experimental and theoretical studies have been made regarding the structural properties and formation of metal nanoparticles. The structure of a nanoparticle depends on the method of synthesis used as well as on the thermal treatment to which the particle turns out to be subject. Hence, we can infer that it is important to count with studies that allow the understanding of the structural and energy changes that happen in the particle as its temperature is varied. Similar works have studied the melting and freezing of nanoparticles at different geometries [12-14]. The purpose of this work is to contribute in a certain extent to improve the understanding of the thermodynamic characteristics of metallic and bimetallic nanoparticles, emphasizing on the surface phenomenology of the particles, since the practical uses of these systems will depend on the behavior and physical properties of the surface.

In this work, we have chosen nanoparticles with icosahedral geometry as starting configurations, since there is wide experimental evidence that the icosahedral geometry is one of the most energetically geometries favorable for metallic particles at this range of sizes [8]. An icosahedron is a geometric structure with 12 vertices connected by 30 edges forming 20 triangular faces. According to their positions in the particle, different atoms occupy identical geometric sites, called equivalent sites. The atoms located in equivalent sites have the same type and number of neighbors, and are placed at the same distance from the origin [15]. We chose an icosahedral nanoparticle of 561 atoms at 5 different concentrations of gold and palladium, where every concentration remained determined by the election of the equivalent sites where the atoms were placed depending on their atomic species, as can be seen in figure 1. We studied the conditions of the melting and freezing transitions for each of these particles.

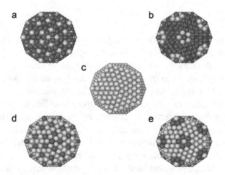

Figure 1. Models that correspond to the initial configurations of the particles with concentrations of (a) $Au_{261}Pd_{300}$, (b)$Au_{271}Pd_{290}$, (c) Au_{561}, (d) $Pd_{261}Au_{300}$ and (e)$Pd_{271}Au_{290}$. The white spheres represent gold atoms, whereas the dark spheres represent palladium atoms.

SIMULATION DETAILS

Using a Sutton-Chen potential model [16] to model the interactions in the Au-Pd bimetallic nanoparticles, we performed the Molecular Dynamics simulations in the canonical ensemble (NVT) making use of the Nosé-Hoover thermostat [17, 18], and following the same strategy used in a previous set of simulations of Au-Pd nanoparticles [19]. Sutton and Chen modified the Finnis-Sinclair potential to study the interactions between particles, and developed it for ten fcc metal species:

$$U(r) = \varepsilon \left[\frac{1}{2} \sum_{j \neq i} \left(\frac{a}{r_{ij}} \right)^n - c \sqrt{\rho_i} \right], \tag{1}$$

where

$$\rho_i = \sum_{j \neq i} \left(\frac{a}{r_{ij}} \right)^m. \tag{2}$$

In the equations 1 and 2, the parameters m and n are positive integers, ε is a parameter with energy dimensions, a is a fcc lattice constant, and c is an adimensional positive parameter that fits according to experimental results.

For every simulation, we used 500,000 time steps, of which 100,000 were used to reach thermal equilibrium and were not considered in the analysis of the results; using a time step of 0.0015 ps we performed a set of NVT simulations, each 750 ps long, for every nanoparticle. For every particle, we performed two series of simulations. In the first series, corresponding to the heating process, we covered a range of temperature from 300 to 1200 K, increasing T every 20 K, using as the initial configuration for every temperature, the final configuration obtained at a previous temperature. In the second series, or cooling process, the swept on temperature was made backing down in temperature from 1200 to 300 K.

To identify the nanoparticles we use the notation $Au_X Pd_Y$ to refer to a particle of X gold atoms and Y palladium atoms.

RESULTS

Structural phase transitions were recognized for every particle at both heating and cooling processes, by analyzing the evolution of the energy during the simulation, and by the analysis of the structure of the particles. The melting temperature for Au_{561} was 660K ±10K, 940K ±10K for $Au_{271}Pd_{290}$, 900K ±10K for $Au_{261}Pd_{300}$, 940K ±10K for $Pd_{271}Au_{290}$ and 960K ±10K for $Pd_{261}Au_{300}$. The melting and freezing temperatures can be identified by the abrupt changes in the slope of the caloric curves, as can be seen in figure 2. We found that for the $Au_{261}Pd_{300}$ and $Au_{271}Pd_{290}$ particles, at temperatures between 700K and 800K the energy decreases with the temperature, while for $Pd_{271}Au_{290}$ and $Pd_{261}Au_{300}$ at temperatures greater than 820K ±10K the energy for this particles increases, due to structural changes before at the melting point, mainly at the surface of the particles. Figure 3 shows snapshots of the particles at different temperatures during the heating process.

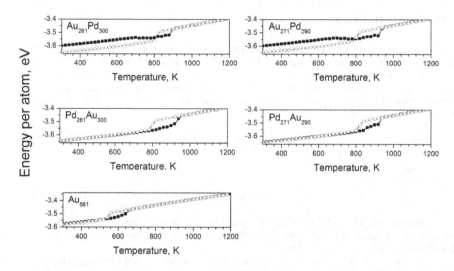

Figure 2. Configuration energy vs. temperature for the particles (a) $Au_{261}Pd_{300}$, (b) $Au_{271}Pd_{290}$, (c)$Pd_{261}Au_{300}$, (d) $Pd_{271}Au_{290}$ and (e) Au_{561}. The dark and light symbols correspond to the heating and cooling processes, respectively.

By the knowledge of the original sites of both atomic species, and analyzing the radial distribution function (RDF), calculated using the center of mass as the origin, we found that in the $Au_{261}Pd_{300}$ and its counterpart $Pd_{261}Au_{300}$, the reversal in the structural role of the atomic species has a very appreciable effect: the $Au_{261}Pd_{300}$ particle has nucleus rich in gold, and its radius is 7Å, while the $Pd_{261}Au_{300}$ has a nucleus mainly made of Pd, and its radius is 6 Å; this remarked difference in the composition of the cores is the main cause that these particles have very pronounced difference in the values of their melting temperatures, since a larger amount of thermal energy is needed for a Pd immersed in the core to migrate out of its original site. This is also the reason why Au atoms have a marked tendency to migrate to the surface of the particle once the thermal energy is enough to allow diffusion. For purposes of comparison, we show in figure 4 the RDF at 300K (quite well before melting), and at a temperature just before melting (between 900 and 960K), when diffusion has already started. Right after melting, for all the particles the composition of gold at the surface becomes larger than at the beginning of the heating process, and larger than the concentration of palladium.

4

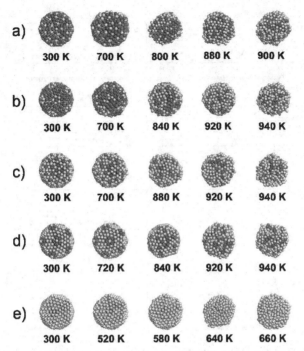

Figure 3. Structural modification of the nanoparticles in the heating process. (a) $Au_{261}Pd_{300}$; (b) $Au_{271}Pd_{290}$; (c) $Pd_{261}Au_{300}$; (d) $Pd_{271}Au_{290}$; and (e) Au_{561}.

To investigate the structural behavior of the particles during the heating and cooling processes, we calculated the global order parameter Q_6 which is defined as [20]

$$Q_6(i) = \left(\frac{4\pi}{2l+1} \sum_{m=-l}^{l} \left| \bar{Q}_{6m}(i) \right|^2 \right)^{\frac{1}{2}},$$

where

$$\bar{Q}_{6m}(i) = \frac{\sum_{i=1}^{N} N_{nb}(i) q_{6m}(i)}{\sum_{i=1}^{N} N_{nb}(i)}.$$

5

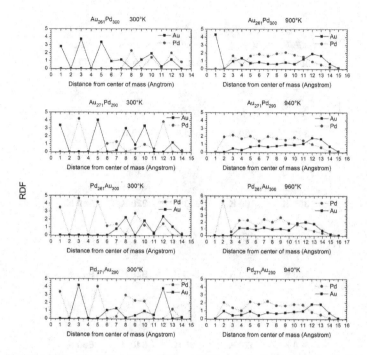

Figure 4. Pd and Au radial distributions, at 300K, and at a temperature just before melting.

N is the total number of atoms in the particle, $N_{nb}(i)$ is the number of first neighbors for the i-th atom, and $q_{6m}(i)$ measures the local order around the i-th atom considering the average of the spherical harmonics Y_{6m} of the bonds with the $N_{nb}(i)$ neighbors:

$$q_{6m}(i) = \frac{\sum_{j=1}^{N_{nb}} Y_{6m}(r_{ij})}{N_{nb}(i)}.$$

Defined in this way, the order parameter Q_6 becomes a fair criteria to identify the resulting structures of the heating process, because a specific structure is characterized by well-defined value of Q_6: a perfect FCC structure gives a value $Q_6^{FCC} = 0.575$, whereas an icosahedron gives $Q_6^{Ih} = 0.663$, while for melted states, the value of Q_6 will drop down to values close to zero. The cutoff distance for identifying the nearest neighbors was taken to be 3.6 A at 300 K. This corresponds to the position of the first minimum in the pair correlation function for fcc Pd-Au. Figure 5 shows the variation in the order parameter with temperature for the nanoparticles.

6

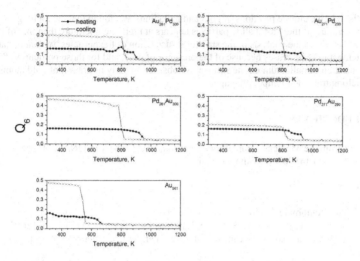

Figure 5. Global order parameter Q_6 vs the temperature for the five different concentrations used in this study.

The final geometries of the particles after the heating-cooling process in general differ from the original configurations, and from the calculation of Q_6 we found that these difference are not just in the shape, but on the overall atomic ordering in the volume of the particle. Only the $Pd_{271}Au_{290}$ and $Au_{261}Pd_{300}$ particles freeze taking an icosahedral shape, while the other particles crystallize as truncated octahedra. The $Pd_{271}Au_{290}$ particle show a larger stability than its $Au_{271}Pd_{290}$ counterpart, that experiences structural changes at temperatures as low as 680K, even when both particles have a very similar melting temperature. Again, this is mostly due to the difference in the amount of Pd between the particles. A similar trend can be noted in a slightly diminished way between the $Au_{261}Pd_{300}$ and the $Pd_{261}Au_{300}$ particles.

CONCLUSION

Gold atoms diffuse towards the surface before the melting temperature for all particles, in agreement with other works [19, 21]. As a result, high concentrations of palladium are not observed in the surface. In the heating curve of the Au_{561} nanoparticle, we see a gradual change in the energy before the melting transition happens. This happens because the atoms in the surface melt before the other ones composing the nanoparticle. For the $Pd_{261}Au_{300}$ and $Pd_{271}Au_{290}$ particles, migration of Au is not necessary in order to minimize the surface energy if gold is present on the surface before thermal equilibrium is. As soon as the heating - cooling cycle is closed, the particle $Au_{290}Pd_{271}$ returns to its original values of potential energy at room temperature, which can be understood as an evidence that the icosahedral geometry is energetically suited for this particle. We observed that the melting transition temperature depends on the relative atomic positions of gold and palladium. We found that the melting transition temperature of the Au-Pd alloy particles appears at higher temperature than that of the pure-gold particle. From the analysis of the structural succession of the particles, we found that the melting temperature increases because of the presence of gold, and for those particles with a higher concentration of palladium on the surface, we observe an early migration of gold atoms before the melting transition temperature appears.

ACKNOWLEDGMENTS

Support from the Mexican Council for Science and Technology (CONACYT, Mexico), through project CIAM 148967, is acknowledged.

REFERENCES

1. O. V. Salata, Nanobiotechnology **2** (3), 1-6 (2004).
2. M. O. Nutt, J. B. Hughes and M. S. Wong, Environ. Sci. Technol. **39** (5) 1346 (2005).
3. I. Bala, S. Hariharan and M.N. Kumar, Crit. Rev. Ther. Drug Carrier Syst. **21** (5), 387 (2004).
4. J.R. Morones, J. L. Elechiguerra, A. Camacho, K. Holt, J. B. Kouri, J. Tapia-Ramírez, and M. José-Yacamán, Nanotechnology **16**, 2346 (2005).
5. M.-L. Wu, D.-H. Chen and T.-C. Huang, Langmuir **17** (13), 3877 (2001).
6. F.-R. Fan, D.-Y. Liu, Y.-F. Wu, S. Duan, Z.-X. Xie, Z.-Y. Jiang, and Z.-Q. Tian. J. Am. Chem. Soc. **130**, 6949 (2008).
7. D.-H. Chen and C.-J. Chen. J. Mater. Chem. **12**, 1557 (2002).
8. R. Ferrando, J. Jellinek and R.L. Johnston, Chem. Rev. **108**, 845 (2008).
9. S. Tajammul Hussain, M. Iqbal and M. Mazhar, J. Nanopart. Res. **11**, 1383 (2008).
10. H. Qian and R. Jin, Nano Lett. **9**, 4083 (2009).
11. M. M. Mariscal, O.A. Oviedo, E.P.M. Leiva, S. Mejía-Rosales and M. José-Yacamán, Nanostructure Science and Technology **3** (26) (2013).
12. J. Shim, B. Lee and Y.W. Cho, Surface Science **512**, 262 (2002).
13. Y. Shibuta and T. Suzuki, Chemical Physics Letters **445**, 265 (2007).
14. Y. Shibuta and T. Suzuki, Chemical Physics Letters **498**, 323 (2010).
15. J.M. Montejano-Carrizales and J.L. Morán-López, Nanostructured Materials **1**, 397 (1992).
16. H. Rafii-Tabar and A.P. Sutton, Philosophical Magazine Letters **63**, 217 (1991).
17. S. Nosé, Molecular Physics **52**, 255 (1984).
18. W.G. Hoover, Phys. Rev. A **31**, 1695 (1985).
19. S. J. Mejía-Rosales, C. Fernández-Navarro, E. Pérez-Tijerina, J. M. Montejano-Carrizales and M. José-Yacamán, J. Phys. Chem. B **110**, 12884 (2006).
20. Y. Chushak and L.S. Bartell, European Physical Journal D **16**, 43 (2001).
21. F Pittaway, L. O. Paz-Borbón, R. L. Johnston, H. Arslan, R. Ferrando, C. Mottet, G. Barcaro, and A. Fortunelli, J. Phys. Chem. C **113**, 9141 (2009).

Mater. Res. Soc. Symp. Proc. Vol. 1479 © 2012 Materials Research Society
DOI: 10.1557/opl.2012.1590

Synthesis and Characterization of Alloys and Bimetallic Nanoparticles of CuNi Prepared by Sol-Gel Method

E.L. de León-Quiroz[1*], D. Vázquez Obregón[1], A. Ponce Pedraza[3], E. Larios-Rodríguez[3], M. José-Yacaman[3] and L.A. García-Cerda[2*]

[1]Instituto Tecnológico de Saltillo, Blvd. V. Carranza No. 2400, Saltillo, Coah., México
[2]Centro de Investigación en Química Aplicada), Blvd. Enrique Reyna No. 140, Saltillo, Coah., C.P. 25294, México
[3]University of Texas at San Antonio. One UTSA Circle. San Antonio, TX. 78249
Corresponding authors: lagarcia@ciqa.mx, yi_lu_lq@hotmail.com

ABSTRACT

Recently, bimetallic nanostructures and nanoalloys have received special interest due to their promising chemical and physical properties. Specifically, Cu-Ni nanoparticles have been investigated for biomedical and catalytic applications. This work reports the synthesis of alloys and bimetallic nanoparticles of Cu_xNi_{100-x} ($x = 20, 40, 50, 60$ and 80) by sol-gel method, and their morphological, structural and magnetic characterization. A precursor material was prepared using a standard Pechini method and then CuNi nanoparticles were obtained by calcination treatments of the precursor in H_2/N_2 atmosphere at 600 and 700°C for 15 minutes. The resulting nanoparticles were characterized by X-ray diffraction (XRD), which reveals that this method led to the formation of CuNi substitutional nanoalloys and bimetallic nanoparticles with good cristallinity related with the calcination temperatures and Cu:Ni weight ratios. Transmission electron microscopy (TEM) shows nearly monodisperse and uniform spherical nanoparticles with sizes between 40 and 70 nm. The magnetic properties were studied using SQUID magnetometry, according with these results, the CuNi nanoparticles showed a ferromagnetic behavior, the magnetization value increases as a function of the weight percentage of Ni.

INTRODUCTION

Nanoparticles are aggregates from a few to many millions of atoms or molecules. These may be formed of the same atoms or molecules, or by two or more different species, these can be studied in different media, such as vapor, colloidal suspensions, or isolated in different matrices [1,2]. The interest in nanoparticles arises in part because they are a new type of material which may have different properties from individual atoms and molecules or bulk material. A major reason for this interest is the evolution of their size-dependent properties and structure [2]. In fact, both the geometry and the energy stability of the groups may change dramatically with size. From the point of view of applications, there is a continuing interest in these materials due to their potential applications in fields such as catalysis and nanoelectronics [2]. Moreover, the range of properties can be expanded to metallic systems if these are mixed with other elements to generate intermetallic compounds and/or alloys. In many cases, there is an improvement in the specific properties of the metallic alloys due to synergistic effects and a rich variety of compositions, structures and properties which give rise to their application in electronics, engineering and catalysis.

Different methods have been used to prepare nanoparticles of CuNi, among which are included the sol-gel [3], mechanical alloying [4], electrochemical process [5], hydrothermal method [6] and microemulsion [7]. The solgel based Pechini method [8] is a wet technique that has several advantages over other methods (mechanical or chemical). This process is simple, inexpensive and very versatile to obtain materials with excellent homogeneity and controlled composition at low temperatures. The general idea of this method is the uniform distribution of metal cations in a polymeric resin precursor, which inhibits segregation and precipitation within the system. The subsequent calcination of the precursor results in obtaining homogeneous nanoparticles at low temperature.

This paper reports the results of the synthesis and characterization of alloys and bimetallic nanoparticles of CuNi obtained by the solgel based Pechini method. Precursor materials were prepared with different molar concentrations of Ni and Cu. The nanoparticles were obtained by calcination treatments at temperatures of 600 and 700 °C. The influence of the molar concentration of the metal ions and the calcination temperature on the structural and morphological properties of the materials was studied. The obtained materials were characterized by X-ray diffraction (XRD), scanning electron microscopy (SEM), transmission electron microscopy (TEM) and SQUID magnetometry.

EXPERIMENT

The analitical grade nickel chloride ($NiCl_2.6H_2O$), copper chloride ($CuCl_2$), citric acid ($C_6H_8O_7.H_2O$) and ethylene glycol ($C_2H_6O_2$) were supplied by Aldrich. All chemicals were used as received. The alloys and bimetallic nanoparticles of CuNi were prepared by solgel based Pechini method as follows. First, the appropriate amounts of citric acid, nickel chloride and copper chloride were mixed and stirred at room temperature until a transparent solution is obtained. Then, ethylene glycol was added and the solution was further stirred to achieve complete solubility. The resulting solution was heated at 130 °C for 24 h to obtain the dry precursor by evaporating the solvent, and then it was ground and calcinated in $10\%H_2/90\%N_2$ atmosphere at 600 and 700 °C for 15 min.

The structure, lattice parameters and phase purity of the synthesized CuNi alloys and bimetallic nanoparticles were determined by X-ray diffraction (XRD) using a Rigaku Ultima IV diffractometer with $CuK\alpha$ radiation source, operated at 25 Kv and 35 mA. Scan was performed 20 to 80° in steps of 0.02° in 2θ scale. The morphology and size of the CuNi materials were studied by scanning electron microscopy (Hitachi S-5500 microscope) and high resolution transmission electron microscopy (HRTEM) using ARMTEM-Jeol microscope. The magnetic properties of the materials were determined by SQUID magnetometry. The chemical composition of the alloys was determined using atomic absorption spectroscopy (AAS) (Varian SpectrAA 250).

RESULTS AND DISCUSSION

Figure 1a shows the X-ray diffraction patterns for the sample obtained at 600 °C for 15 min under a reducing atmosphere composed of H_2/N_2. In the figure can be observed reflections representative of fcc structure with the corresponding (111) and (200) planes. The angles of these reflections in all the samples were greater than those typical for pure Cu (43.3 and 50.4°) and lower than those for pure Ni (44.5 and 51.8°) according to the standards JCPDS 04-0836 for

10

copper and 04-0850 for nickel. The shift of peaks positions confirmed the formation of a solid solution or alloy between Cu and Ni. The lattice parameter (a) obtained from the respective d-values corresponding to the (111) diffraction peaks of the alloys are shown in the Table I, for all the samples the lattice parameter value is intermediate between the lattice parameters of pure Cu (0.3614 nm) and Ni (0.3524 nm). It is well known that the ionic radii of Cu and Ni are 0.73 Å and 0.69 Å respectively. In the CuNi alloys, the Cu cation is bigger than the Ni cation and when it replaces the Ni site, the crystal lattice will expand. The expansion of the crystal lattice has been observed in CuNi alloys obtained by different methods [4, 9]. The crystallite sizes of the CuNi alloys were calculated from the XRD patterns according to the linewidth of the most intense diffraction peak (111) using Scherrer Equation [10]. The crystallite sizes of CuNi nanoalloys are presented in Table I, the sizes vary with the composition of each sample in values below 20 nm.

For the samples obtained at 700 °C, the XRD diffraction patterns are shown in the Figure 2b. The X-ray reflections of these samples clearly indicate the presence of two distinct materials: a copper-rich phase and a nickel-rich phase. Again the peaks positions in the samples are slightly shifted from those of pure copper and nickel. The lattice parameter value for the Cu-rich and Ni-rich phases is intermediate between the pure metallic phases.

There is a clear evolution of the diffraction reflections as the calcination temperature increases, showing a gradual splitting of a single-reflection pattern at 600 °C into a two-reflections overlapping pattern upon increasing the temperature to 700 °C. The mechanism for to obtain alloys and bimetallic nanoparticles is not clear at this moment. The literature mentions that depending of the temperature, differences in the surface free energy of the metals can favor the formation of separated phases [11].

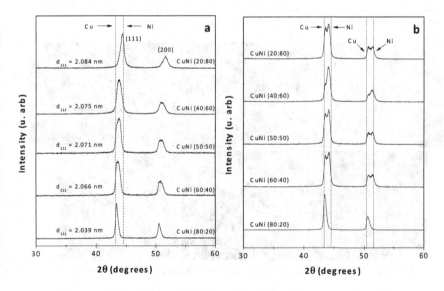

Figure 1. XRD patterns of CuNi alloys (a) and bimetallic nanoparticles (b).

The weight percentages of nickel and copper obtained by AAS for CuNi samples are shown in Table I. The chemical composition was very close with the theorical composition used for the preparation of the samples.

Table I. Interplanar distance, lattice parameter and chemical composition of CuNi alloys.

Sample	Interplanar distance, $d_{(111)}$ (nm)	Lattice parameter, a (Å)	Chemical composition (AAS)	
			Cu	Ni
Cu80Ni20	2.0839	3.6092	83.7	16.3
Cu60Ni40	2.0751	3.6028	58.8	41.2
Cu50Ni50	2.0712	3.5731	54.3	45.7
Cu40Ni60	2.0657	3.5693	45.6	54.4
Cu20Ni80	2.0395	3.5539	18.5	81.5

Figure 2 shows the SEM images of samples Cu20Ni80 (a, d), Cu50Ni50 (b, e) and Cu80Ni20 (c, f) obtained at 600 and 700° C, respectively. At 600 °C, the particles have spherical shape with sizes between 40 and 70 nm, this morphology is presented in CuNi particles prepared by microemulsion [9]. At 700 °C, a mixture of spherical particles and rods is present. Generally, the particles are embedded in a carbon matrix, this behavior is related to two factors: firstly, the calcination time does not allow the total decomposition of organic material and secondly associated with the use of the reducing atmosphere, which displaces oxygen, avoiding the total burning of the precursor material, this behavior has been observed in the preparation of nanoparticles embedded in a silicate matrix by solgel [12].

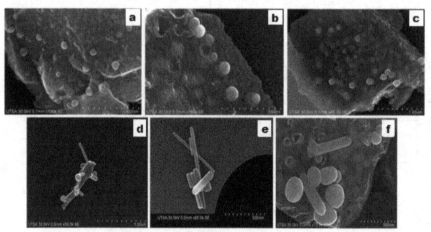

Figure 2. SEM images for CuNi alloys (a-c) and bimetallic nanoparticles (d-f).

The typical TEM image of the CuNi alloys is shown in Figure 3a. The nanoparticles are spherical in shape, with sizes below 100 nm, corroborating the observations of SEM. The high-resolution TEM image (Figure 3b) shows the lattice fringes and indicates the high crystallinity of the CuNi alloys, the distance between two fringes was estimated to be 2.07 Å consistent with d_{111} reflection of a fcc structure.

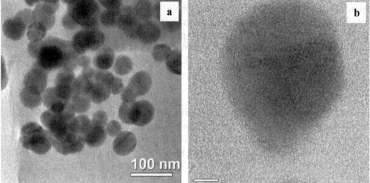

Figure 3. TEM (a) and HRTEM (b) images of Cu50Ni50 alloy.

Figure 4 shows hysteresis loops of CuNi alloys obtained at 600 °C. The Cu80Ni20 sample was found to be paramagnetic in nature. The samples Cu50Ni50 and Cu20Ni80 exhibit ferromagnetic behavior. The maxima saturation magnetization (Ms) corresponds to the sample with larger content of nickel (11.8 emu/g), which is responsible for the magnetic behavior of the system.

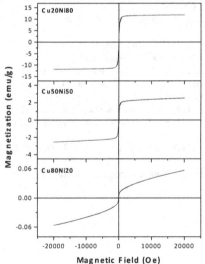

Figure 4. Hysteresis loops of CuNi alloys.

CONCLUSIONS

In this work alloys and bimetallic nanoparticles of CuNi were prepared using a solgel based Pechini method. Through changes in the concentration of metal ions and calcination temperatures, the percentage of the phases present in the samples was controlled. SEM and HRTEM studies showed that it is possible to obtain particles with spherical and/or rod morphology. The proposed route can be extended to other metals such as Fe, Cu and Co.

ACKNOWLEDGMENTS

The authors acknowledge financial support from CONACYT grant number 133991. E.L. de Leon-Quiroz acknowledges CONACYT for the scholarship No. 371542 and the economic support for her stay as graduate research scholar at The University of Texas at San Antonio under the program "becas mixtas". Also the authors thank to J.A. Espinoza Muñoz for the AAS analysis; Dario Bueno and Gilberto Hurtado for the VSM measurements.

REFERENCES

1. J. Jellinek, "Theory of Atomic and Molecular Clusters", (Springer, 1999).
2. R.L. Johnston, "Atomic and Molecular Clusters", (Taylor and Francis, 2002).
3. S. Pramanik, S. Pal, De. G. Bysakh, *J. Nanopart. Res.* **13**, 321 (2011).
4. L. Durivault, O. Brylev, D. Reyter, M. Sarrazin, D. Bélanger, L. Roué, *J. Alloys Compd.* **432**, 323 (2007).
5. R. Qiu, X.L. Zhang, R. Qiao, Y. Li, Y.I. Kim, Y.S. Kang, *Chem. Mater.* **19**, 4174 (2007).
6. G. H. Mohamed Saeed, S. Radiman, S.S. Gasaymeh, H.N. Lim, N.M. Huang, *J. Nanomater.* **Article ID 184137**, 5 pages (2010).
7. J. Feng, C.H. Ping Zhang, *J. Colloid. Interf. Sci.* **293**, 414 (2006).
8. Maggio P. Pechini, U.S. Patent No. 3 330 697 (11 July 1967).
9. J. Ahmed, K.V. Ramanujachary, S.E. Lofland, A. Furiato, G. Gupta, S.M. Shivaprasad, A.K. Ganguli, *Colloid Surface A* **331**, 206 (2008) .
10. B.D. Cullity BD, S.R. Stock, "Elements of X-ray diffraction, 3rd ed. (Prentice Hall, 2001).
11. B.N. Wanjala, J. Luo, B. Fang, D. Mott, C.J. Zhong, *J. Mater. Chem.*, **21**, 4012 (2011).
12. G. Maduraiveeran; P. Manivasakan; R. Ramaraj, *Int. J. of Nanotechnology,* **8**, 925 (2011).

Mater. Res. Soc. Symp. Proc. Vol. 1479 © 2012 Materials Research Society
DOI: 10.1557/opl.2012.1591

Stability, Electronic Properties, and Structural Isomerism in Small Copper Clusters

Juan M. Montejano-Carrizales, Faustino Aguilera-Granja, and Ricardo A. Guirado-López
Instituto de Física, Universidad Autónoma de San Luis Potosí
San Luis Potosí, 78140 S.L.P., MEXICO

ABSTRACT

We present extensive pseudopotential density functional theory calculations dedicated to analyze the stability, electronic properties, and structural isomerism in Cu_6 clusters. We consider structures of different symmetries and charge states. Our total energy calculations reveal a strong competition between two- and three-dimensional atomic arrays, the later being mostly energetically preferred for the anionic structures. The bond lengths and electronic spectra strongly depend on the local atomic environment, a result that is expected to strongly influence the catalytic activity of our clusters. Using the nudged elastic band method we analyze the interconversion processes between different Cu_6 isomers. Complex atomic relaxations are obtained when we study the transition between different cluster structures; however relatively small energy barriers of approximately 0.3 eV accompany the atomic displacements. Interestingly, we obtain that by considering positively charged Cu_6^+ systems we reduce further the energy barriers opposing the interconversion process. The previous results could imply that, under a range of experimental conditions, it should be possible to observe different Cu_6 cluster structures in varying proportions.
Keywords: electronic structure, Cu, nanostructures.

INTRODUCTION

The structural properties and electronic behavior of small copper clusters have been of particular interest due to their promising applications in catalysis, optoelectronics, and nanophotonics [1]. In general, the size and atomic structure of cluster structures are of fundamental importance to determine the overall characteristics and their response to external fields (see Ref. 1 and papers therein). However, as it is well known, direct experimental determination of the cluster geometry remains extremely challenging [2]. Interestingly, indirect experimental evidence of the structure of a given cluster, or even information about the presence of various structural isomers in the samples [3], can be obtained by means of ion mobility experiments[4] and reactivity studies [3]. With the previous two types of experiments, a more clear understanding of the measured data (e.g., optical absorption spectra) has been obtained.

From the theoretical point of view, various density functional theory (DFT) calculations have been performed to analyze the structure, electronic properties, and stability of small Cu clusters [2]. With respect to the determination of the lowest energy atomic configuration for a given cluster size the published data still remains controversial. However, it has been clearly established that i) there is a strong competition between planar (2D) and three-dimensional (3D) atomic configurations, ii) the calculations are found to be sensitive to the nature of the exchange-correlation potential and basis set. Furthermore, it is clear that the most favored isomer in energetic terms and zero temperature are not expected to be the cluster structure present in the experiments and, as a consequence, the understanding of the electronic properties of small Cu clusters is still very challenging.

At this point we would like to comment that several experimental measurements on small clusters have been interpreted by assuming the co-existence of various isomers in the samples[3]. As a consequence, not only theoretical reports dedicated to analyze minimum energy structures are important but also studies addressing the interconversion processes between close-in-energy isomers are highly desirable. In this work, we present thus an extensive pseudopotential density functional theory set of calculations dedicated to analyze the stability, electronic properties, and structural isomerism in Cu_6 clusters. Our total energy calculations show that there is a strong competition between two- and 3D atomic arrays. Using the nudged elastic band (NEB) method[5] we analyze the interconversion processes between different Cu_6 clusters. Complex atomic relaxations are obtained; however we will show that relatively small energy barriers accompany them. The previous results could imply that, under a range of experimental conditions, it should be possible to observe different Cu_6 cluster structures in varying proportions.

The rest of the paper is organized as follows. In Sec. II we present our theoretical methodology. In Sec. III we discuss our theoretical results and, finally, in Sec. IV the summary and conclusions are given.

THEORETICAL METHODOLOGY

The structural and electronic properties of our considered Cu_6 clusters will be obtained within the DFT approach using the ultrasoft pseudopotential approximation for the electron-ion interaction and a plane wave basis set for the wavefunctions with the use of the PWscf code [6]. For all our considered structures, the cutoff energy for the plane wave expansion is taken to be 408 eV. A cubic supercell with 30 Å size was employed in the calculations and the Γ point for the Brillouin zone integration. In all cases, we have used the Perdew-Burke-Ernzerhof (PBE) pseudopotential [7] and perform fully unconstrained structural optimizations for all our considered isomers using the conjugate gradient method. The convergence in energy was set as 1 meV and the structural optimization until a value of less than 1 meV/Å was achieved for the remaining forces for each atom.

To determine the minimum energy paths (in a static approximation) as well as the transition states, we applied the nudged elastic band (NEB) methodology [6]. The NEB calculation scheme is a chain-of-states method where a set of images between the initial and final states must be created to achieve a smooth curve. In our particular case, the small size of our copper clusters will ensure that the calculations will remain computationally tractable. In all our calculations, we use nine images to determine the energy profiles, which have been found to be enough to reveal the different stages of the interconversion processes in our here-considered Cu_6 structures. The relevant energy barriers between well-defined initial and final atomic configurations were obtained by calculating the energy difference of the initial position and the saddle point of each one of the energy profiles.

RESULTS

In Figure 1 we show our optimized atomic configurations for the Cu_6 isomers considered. We analyze both planar and three-dimensional clusters and study the relative stability between the two types of structures that has been the subject of controversy in the literature [2]. In the figure we order the structures from the most stable [Figure 1(a)] to the less stable [Figure 1(h)]

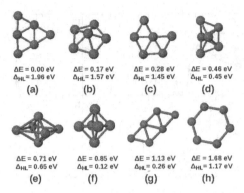

ΔE = 0.00 eV ΔE = 0.17 eV ΔE = 0.28 eV ΔE = 0.46 eV
Δ_{HL}= 1.96 eV Δ_{HL}= 1.57 eV Δ_{HL}= 1.45 eV Δ_{HL}= 0.45 eV
(a)   (b)   (c)   (d)

ΔE = 0.71 eV ΔE = 0.85 eV ΔE = 1.13 eV ΔE = 1.68 eV
Δ_{HL}= 0.65 eV Δ_{HL}= 0.12 eV Δ_{HL}= 0.26 eV Δ_{HL}= 1.17 eV
(e)   (f)   (g)   (h)

Figure 1. Optimized atomic configurations for our considered Cu_6 isomers. We also include the energy difference between the different isomers with the respect to the more stable one [(a)]. In addition, we specify the energy difference of the HOMO—LUMO gap for each one of the clusters.

atomic arrays. From our calculations we have found that the lowest energy Cu_6 cluster is defined by the planar structure shown in Figure 1(a) followed by the three-dimensional pentagonal pyramid presented in Figure 1(b). The energy difference between the two isomers is of 0.17 eV. Notice that even the planar structure shown in Figure 1(c) is more stable than the well known compact arrays shown in Figures 1(d)—1(f) by values of the order of 0.2—0.6 eV. The previous results clearly illustrate the tendency of very small Cu clusters to adopt 2D configurations. In all cases, the Cu-Cu bond lengths vary in the range of 2.38—2.60 Å and the energy difference between the highest occupied and lowest unoccupied molecular orbitals (so called HOMO-LUMO gap) goes from 0.12 to 1.96 eV, being the copper isomer with the largest electronic gap the most stable Cu_6 cluster. Notice that the large variations exhibited by the HOMO-LUMO gap in our copper clusters are very interesting from the point of view of their catalytic activity since structures with reduced gaps are expected to be more reactive. Furthermore, the absorption spectra is expected to be also very sensitive to these variations, in particular the optical gap, a result that could be very useful to provide finger prints to identify the possible structure adopted by copper clusters of these sizes.

In Figure 2, we present the average density of states (ADOS) of the most stable atomic configurations shown in Figures 1(a) and 1(b). When comparing Figures 2(a) and 2(b) we see that the electronic spectra of the planar structure is more degenerated than the one found for the three-dimensional Cu_6 cluster. Notice that the reduced coordination number of the former is reflected in the existence of a smaller occupied band width [see Figure 2(a)] and that, as already mentioned above, a largest energy separation between the HOMO and LUMO levels is obtained. Finally, we must comment that it would be very interesting to compare all of our calculated ADOS with available photoelectron spectroscopy experiments for clusters of these sizes in order to gain more information related with the possible atomic configuration adopted by these very small particles.

In Figure 3, we present our results for the interconversion process between our two most favorable neutral isomers shown in Figures 1(a) and 1(b). In our calculations we use a total of

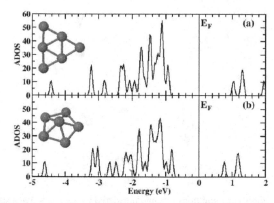

Figure 2. The average density of states (ADOS) of the two most stable atomic configurations for Cu$_6$ clusters shown in Figures 1(a) and 1(b). In the inset we show the atomic configurations of the considered isomers.

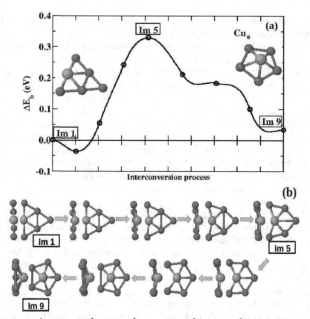

Figure 3. Interconversion process between the two neutral isomers shown in Figures 1(a) and 1(b). In (a) we plot the energetic associated with this process and in (b)we show the sequence of atomic relaxations involved in the structural transformation. The gray atom corresponds to a Cu species defined as a reference point along the path.

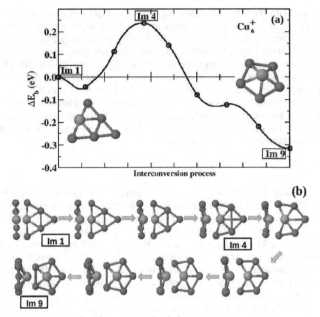

Figure 4. Same as in Figure 3 but for cationic Cu_6^+ clusters.

nine images to construct the reaction pathway that have been found to be enough to reveal the sequence of atomic displacement and rotations leading to the interconversion between the two considered Cu_6 clusters. We define a reference species along the path as the gray atom included in the clusters. In Figure 3(a) we show the energetics associated with the previous process and, in Figure 3(b), we present the sequence of structural transformations [showing both the side (left) and front (right) views] occurring along the path. From Figure 3(a) we see that an energy barrier (direct barrier) of the order of 0.35 eV separates both planar and three-dimensional isomers. In contrast, when moving from the pentagonal pyramid to the planar atomic array, a smaller (reverse) barrier of 0.3 eV needs to be overcome. Remark that, at the transition state (Im 5), the reference atom is already bonded to the rest of the atoms of the Cu_6 cluster with more contracted Cu-Cu distances that vary in the range of 2.35—2.40 Å. Notice also that some three-dimensional character is observed since our reference Cu atom not only performs lateral relaxations but also notable upward displacements. After Im 5, the system spontaneously (with no barrier) moves out of the plane of the Cu_5 ring ending up finally in an on-hollow adsorbed configuration.

In real experiments, cluster systems are found as charged species and, as a consequence, it is important to analyze again the interconversion process shown in Figure 3 but for positively charged Cu_6^+ clusters. In Figure 4 we show thus the energetics [Figure 4(a)] and sequence of structural transformations [Figure 4(b)] joining the structures shown in Figures 1(a) and 1(b) but as cationic species. From Figure 4(a) we note that the energy barrier required to go from the planar structure to the 3D cluster is reduced, when compared to the one found in Figure 3(a), being now of 0.25 eV. This result is very interesting and could imply that new (and easier) reaction channels can be open by a charging mechanism. It is also important to note that, in the positively charged species, the 3D Cu_6^+ is more stable than the planar array. As a consequence, extracting electrons from our cluster systems can also alter the abundance of different isomers.

Finally, notice that even if the direct and reverse barriers are strongly modified by charging i) the sequence of structural transformations remains almost the same and ii) the transition state in Figure 4(b) has a similar atomic structure than in the case of the interconversion process for neutral clusters. We believe our results shed some light in to the structural isomerism of small copper clusters and imply that the experimental measurements performed in this kind of systems could be the result of a complex average of the properties of different isomers simultaneously present in the samples.

CONCLUSIONS

We have presented an extensive pseudopotential density functional theory calculations devoted to analyze the stability, electronic properties, and structural isomerism in Cu_6 clusters. We consider various cluster structures having different symmetries and charge states already considered by previous authors. We found a strong competition between two- and three-dimensional atomic arrays. The bond lengths and electronic spectra strongly depend on the local atomic environment, a fact that could play an important tole in the catalytic activity of our clusters. The interconversion processes between different Cu_6 isomers reveal that, even if complex atomic relaxations are obtained, relatively small energy barriers of approximately 0.3 eV accompany the atomic displacements. Interestingly, we obtain that for positively charged Cu_6^+ systems the energy barriers opposing the interconversion process are reduced. The previous results could imply that, under a range of experimental conditions, it should be possible to observe different Cu_6 cluster structures in varying proportions.

ACKNOWLEDGEMENTS

Authors acknowledge partyal financial support from PROMEP-SEP-Mexico, CONACyT 162651 and 169345, and to the Ministerio de Educacion, Cultura y Deporte, Ref. SAB2011-0024, Spain.

REFERENCES

1. K. Baishya, J. C. Idrobo, S. Öğüt, M. Yang, K. A. Jackson, and J. Jellinek, Phys. Rev. B **83**, 245402 (2011).
2. G. Guzman-Ramirez, F. Aguilera-Granja, and J. Robles, Eur. Phys. Jour. D **57**, 49 (2010).
3. M. S. Ford, M. L. Anderson, M. P. Barrow, D. P. Woodruff, T. Drewello, P. .J. Derrick, and S. R. Mackenzie, Phys. Chem. Chem. Phys. **7**, 2005, 975 (2005).
4. A. K. Starace, C. M. Neal, B. Cao, M. F. Jarrold, A. Aguado, and J. M. López, J. Chem. Phys. **131**, 044307 (2009).
5. G. Henkelman, B. P. Uberuaga, and H. Jonsson, J. Chem. Phys. **113**, 9901 (2000).
6. S. Baroni, A. Dal Corso, S. de Gironcoli, P. Giannozzi, C. Cavazzoni, G. Ballabio, S. Scandolo, G. Chiarotti, P. Focher, A. Pasquarello, K. Laasonen, A. Trave, R. Car, N. Marzani, and A. Kokalj, http://www.pwscf.org/.
7. J. P. Perdew, K. Burke, and M. Ernzerhof, Phys. Rev. Lett. **77**, 3865 (1996).

Mater. Res. Soc. Symp. Proc. Vol. 1479 © 2012 Materials Research Society
DOI: 10.1557/opl.2012.1592

Ultrasonic Milling and Dispersing Technology for Nano-Particles

Kathrin Hielscher[1]
[1]Hielscher Ultrasonics GmbH, Warthestr. 21, 14513 Teltow, Germany
kathrin@hielscher.com; _www.hielscher.com_

ABSTRACT

Ultrasonically generated forces are well known for dispersing and deagglomeration of small volumes in laboratory and bench-top scale. By the evaluation and optimization of the most important ultrasonic parameters and the development of large scale ultrasonic machinery, ultrasound forces can be applied also for particle size reduction and wet-milling of nano-particles in industrial scale.

Keywords: acoustic, nanoscale, nano-structure, ultrasound, dispersion, particle size reduction, ink

INTRODUCTION

The dispersing and deagglomeration of solids into liquids is an important application of ultrasonic devices. If powders are wetted, the individual particles build agglomerates and are held together by attraction forces of various physical and chemical natures, including van der Waals forces and liquid surface tension. This effect is stronger for higher viscosity liquids, such as polymers or resins. The attraction forces must be overcome on order to deagglomerate and disperse the particles into liquid media. An even dispersion and deagglomeration is important to use the full potential of the particles. Especially nano-particles offer extraordinary characteristics, which can only be exploited in highly even dispersed state.

The application of mechanical stress – e.g. generated by ultrasonic cavitation - breaks the particle agglomerates apart. Also, liquid is pressed between the particles. Different technologies are commonly used for the dispersing of powders into liquids. This includes high pressure homogenizers, agitator bead mills, impinging jet mills and rotor-stator-mixers.

High intensity ultrasonication is an interesting alternative to these technologies and particularly for the particle treatment in the nano-size range the only effectual method to achieve the required results.

ULTRASONIC CAVITATION

By high-power/ low-frequency ultrasound high amplitudes can be generated. Thereby, high-power/ low-frequency ultrasound can be used for the processing of liquids such as mixing, emulsifying, dispersing and deagglomeration, or milling. When sonicating liquids at high intensities, the sound waves that propagate into the liquid media result in alternating high-pressure (compression) and low-pressure (rarefaction) cycles, with rates depending on the frequency. During the low-pressure cycle, high-intensity ultrasonic waves create small vacuum bubbles or voids in the liquid. When the bubbles attain a volume at which they can no longer absorb energy, they collapse violently during a high pressure cycle. This phenomenon is termed cavitation. Cavitation, that is "the formation, growth, and implosive collapse of bubbles in a liquid. Cavitational collapse produces intense local heating (~5000 K), high pressures (~1000

atm), and enormous heating and cooling rates (>109 K/sec)" and liquid jet streams (~400 km/h)".
[5]
There are different means to create cavitation, such as by high-pressure nozzles, rotor-stator mixers, or ultrasonic processors. In all those systems the input energy is transformed into friction, turbulences, waves and cavitation. The fraction of the input energy that is transformed into cavitation depends on several factors describing the movement of the cavitation generating equipment in the liquid. The intensity of acceleration is one of the most important factors influencing the efficient transformation of energy into cavitation. Higher acceleration creates higher pressure differences. This in turn increases the probability of the creation of vacuum bubbles instead of the creation of waves propagating through the liquid. Thus, the higher the acceleration the higher is the fraction of the energy that is transformed into cavitation. In case of an ultrasonic transducer, the amplitude of oscillation describes the intensity of acceleration. Higher amplitudes result in a more effective creation of cavitation. In addition to the intensity, the liquid should be accelerated in a way to create minimal losses in terms of turbulences, friction and wave generation. For this, the optimal way is a unilateral direction of movement. This makes ultrasound an effective means for the dispersing and deagglomeration but also for the milling and fine grinding of micron-size and sub micron-size particles. [1]

In addition to its outstanding power conversion, ultrasonication offers the great advantage of full control over the most important parameters: Amplitude, Pressure, Temperature, Viscosity, and Concentration. This offers the possibility to adjust all these parameters with the objective to find the ideal processing parameters for each specific material. This results in higher effectiveness as well as in optimized efficiency.

PARAMETERS OF ULTRASONIC PROCESSING

Ultrasonic liquid processing is described by a number of parameters. Most important are amplitude, pressure, temperature, viscosity, and concentration. The process result, such as particle size, for a given parameter configuration is a function of the energy per processed volume. The function changes with alterations in individual parameters. Furthermore, the actual power output per surface area of the sonotrode of an ultrasonic unit depends on the parameters. The power output per surface area of the sonotrode is the surface intensity (I). The surface intensity depends on the amplitude (A), pressure (p), the reactor volume (VR), the temperature (T), viscosity (η) and others.

$$I[W / mm^2] = (A[\mu m], p[bar], VR [ml], T[°C], \eta[cP],...) .$$
$$\quad\quad\quad + \quad\quad + \quad\quad - \quad\quad - \quad\quad +$$

(+ indicates cavitation-intensifying parameters, - indicates cavitation-reducing parameters)

The impact of the generated cavitation depends on the surface intensity. In the same way, the process result correlates. The total power output of an ultrasonic unit is the product of surface intensity (I) and surface area (S):
$$P [W] = I [W / mm^2] * S[mm^2]$$

For an effective and efficient sonication, these parameters have to been established and optimized. [1]

DEMONSTRATION OF ULTRASONIC MILLING AND DISPERSING EFFECTS BY MEANS OF SPECIFIC MATERIAL EXAMPLES

In the ink, paint and coating industries the dispersing, deagglomeration, and wet-milling of pigment powders and nano particles is a basic application with fundamental consequence for the product quality. For pigments, the characteristic of changing the color of reflected or transmitted light as the result of wave length-absorption, makes organic and inorganic pigments the important raw material in the production of inks, printing colors, varnishes, coatings etc. The majority of pigments is metallic and bases on titanium, iron oxide, zinc, copper, cadmium, cobalt, mercury, ultramarine, lead, chromium, and clay earth. Carbon is another important material of ink and paint pigments.

As pigments are hard and insoluble, higher efforts during the processing are required. High-performance pigments provide an intense color-strength, good gloss, transparency, light fastness, weather fastness, resistance against heat, moisture, as well as chemical resistance, stability during processing and transfer efficiency.

Most pigments are very cost-intensive – consequently, the endeavor of the ink manufacturing industry is focused on production of high-quality ink with the most color strength from the least possible amount of pigment. A faster dispersion and an increase of the production capacity help reducing the costs. The grade of the uniformity of particles and the evenness of the dispersion are essential to achieve higher ink gloss, more intensive color strength and a better overall appearance. [3]

Due to these facts, reliable high-power equipment for pigment processing is needed. Ultrasonic milling and dispersing processors tested in different studies for their efficiency and reliability in regards to their performance of pigment treatments. For the ink manufacture, pigments have to be finely grounded and dispersed into a liquid medium. The requirements, that have to be fulfilled by pigment dispersions, include a particle size less than 150nm, colloidal stability, the compatibility of various ink components, and purity.

Ultrasonic milling and dispersing is a well-known and proven technology to achieve small particles with particle sizes in the range from 500µm down to approx. 10nm.

Ultrasonic Milling of Pigments for Inks

The images below show the result of ultrasonic dispersion of carbon black pigment in UV ink. The particle size reduction and the even dispersion grade are conspicuous.

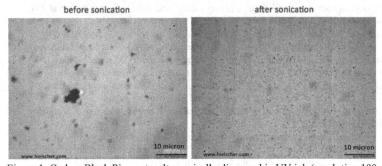

Figure 1: Carbon Black Pigments, ultrasonically dispersed in UV ink (resolution 100x)

As the particle size is determinant for color strength, surface finish, and influences the delivery method, finally the dispersion is decisive for the ink quality. The stability of the ink, its long-term performance and rheology are results of dispersion technique, too.
The microscope pictures below show the milling effect of ultrasound, using oil-based magenta pigment.

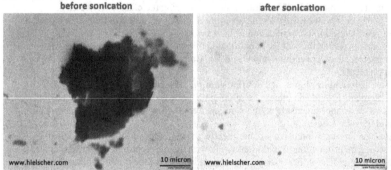

Figure 2: Microscope progress images of ultrasonic milling of oil-based magenta pigment (resolution 100x)

INDUSTRIAL IMPLEMENTATION OF ULTRASOUND

Ultrasonic processing of particles allows processing all particles evenly. Hielscher's industrial ultrasonic processors are commonly used for inline sonication. Therefore, the suspension is pumped into the ultrasonic reactor vessel. There it is exposed to ultrasonic cavitation at a controlled intensity. The exposure time is a result of the reactor volume and the material feed rate. Inline-sonication eliminates by-passing because all particles pass the reactor chamber following a defined path. As all particles are exposed to identical sonication parameters for the same time during each cycle, ultrasonication typically shifts the distribution curve rather than widening it. Generally, "right tailing" cannot be observed at sonicated samples. The option of repeated ultrasonic processing by a loop setup allows to find the perfect sonication for every pigment and every ink formulation. Such treated pigment particles result in better ink quality and show higher stability, an increased self life (also at elevated temperatures), freeze-thaw stability, reduced flocculation stable rheology and lower viscosity at higher particle loading.
High power equipment uses more electricity. Considering rising energy prices, this affects the costs of processing. For this reason, it is important, that the equipment does not lose much energy in the conversion of electricity into mechanical output. Regarding energy consumption, ultrasound is to name as very energy efficient. Hielscher's ultrasonic devices convince by outstanding performance and high efficiency. The ultrasonic processors have an outstanding efficiency of >85%. This reduces your electricity costs and gives you more processing performance. Kusters et al. (1994) sum up in their study that ultrasonic fragmentation is equally efficient as conventional grinding. [2]
In another study, Pohl et al. compared the processing efficiency of ultrasonic dispersion of silica with other high-shear mixing methods, such as with an IKA Ultra-Turrax (rotor-stator-system). Pohl et al. compared the particle size reduction of Aerosil 90 (2%wt) in water using an Ultra-

Turrax (rotor-stator-system) at various settings with that of a Hielscher UIP1000hd ultrasonic device in continuous mode. The table below shows the results.

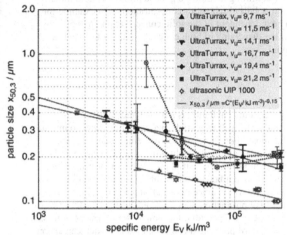

Figure 3: Comparison of particle size reduction of Aerosil90 (2%wt) by an Ultra-Turrax and by a Hielscher ultrasonic device [4]

The study of Pohl et al. concludes that *"at constant specific energy E_V ultrasound is more effective than the rotor-stator-system"* and that *"the applied ultrasound frequency in the range from 20 kHz up to 30 kHz has no major effect on the dispersion process."* [4]

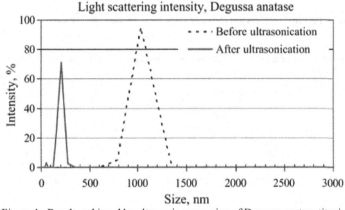

Figure 4 - Results achieved by ultrasonic processing of Degussa anatase titanium

The breakup of the agglomerate structures in aqueous and non-aqueous suspensions allows utilizing the full potential of nano-size materials. Investigations at various dispersions of nanoparticulate agglomerates with a variable solid content have demonstrated the considerable advantage of ultrasound when compared with other technologies, such as rotor stator mixers (e.g. ultra turrax), piston homogenizers, or wet milling methods, e.g. bead mills or colloid mills. Hielscher ultrasonic systems can be run at fairly high solids concentrations. For example for silica the breakage rate was found to be independent of the solid concentration up to 50% by weight. Ultrasound can be applied for the dispersing of high concentration master-batches – processing low and high viscosity liquids.

CONCLUSION

Dispersing and wet-milling by ultrasonic cavitation is a proven technique to achieve evenly distributed dispersions at nano-range as well as particle size reduction down to micron- and nano-size. The full control over the most important parameters – amplitude, pressure, temperature, viscosity, concentration – allows finding the right process adjustment regarding particle characteristics and aimed size. By ultrasonic industrial processors in the power range between 500 watts up to 16 kilowatts per device it becomes possible to develop specific process setups to fulfill specific requirements. By the wide device range, all steps of development –from first testing to process optimization and final production – are covered. The advantages shown above turn ultrasonic dispersing and milling into a potential technology for industrial processing in various sectors, such as for the production of paints and coatings, ink and inkjet, cement and concrete, or cosmetics.

REFERENCES

1. Th. Hielscher, in *Proceedings of European Nanosystems Conference ENS* (2005) pp. 138-143.
2. K. A Kusters, S. E. Pratsinis, S. G. Thomas, and D. M. Smith: Powder Technol. **80**, 253 (1994).
3. A. Pekarovicova, and J. Pekarovic, *Emerging Pigment Dispersion Technologies*. Pira International (2009).
4. M. Pohl and H. Schubert, PARTEC 2004.
5. K. S. Suslick, in *Kirk-Othmer Encyclopedia of Chemical Technology*, 4th ed. (1998) pp. 517-541.
6. www.hielscher.com/mill

Mater. Res. Soc. Symp. Proc. Vol. 1479 © 2012 Materials Research Society
DOI: 10.1557/opl.2012.1593

A Portable Setup for Molecular Detection by Transmission LSPR

Giulia Cappi[1], Enrico Accastelli[1], Fabio M. Spiga[1], Vera Cantale[2], Maria A. Rampi[2], Luca Benini[3], Carlotta Guiducci[1].
[1]Laboratory of Life Sciences Electronics, Swiss Federal Institute of Technology (EPFL), 1015 Lausanne, Switzerland.
[2]Dipartimento di Chimica, Università di Ferrara, via Borsari 46, 44121 Ferrara, Italy.
[3]Dipartimento di Elettronica, Informatica e Sistemistica, Università di Bologna, Viale Risorgimento 2, 40136 Bologna, Italy.
Contact author's e-mail: giulia.cappi@epfl.ch

ABSTRACT

In the framework of bioanalytics and multiple array detection, we developed a fully portable and low-cost detection system based on Localized Surface Plasmon Resonance (LSPR) in a transmission configuration (T-LSPR). The transmission approach is suitable to be scaled to small dimension systems and to enable high-density array measurements on the same platform. Our setup is made out of off-the-shelf components and consists of a set of discrete light sources and a couple of light-detectors which enable a differential measurement setup. An algorithm fits the measured data and extracts the information of the plasmon peak position in the spectrum. The performance of our T-LSPR measurement system has been characterized on a set of Fluorinated Tin Oxide-coated glass slides covered with gold Nanoislands (NIs). The samples have been modified with a single-stranded DNA layer and a real-time DNA hybridization experiment has been performed. Here we demonstrate that the proposed T-LSPR device, based on the characterization of the plasmon peak with a differential approach, is able to monitor real-time DNA hybridization on surface, and to precisely measure the position of the peak with a standard deviation in wavelength of 0.2 nm.

Keywords: sensor, biological, optical.

INTRODUCTION

Parallel detection of multiple analytes by means of high-throughput self-contained automated analytical systems is at the basis of future diagnostics. T-LSPR [1-4] is a very sensitive label-free technique suitable for arrays implementation [5]. It is based on the excitation of the interface between a dielectric and a non connected pattern of metal, on which a localized surface plasmon wave arises. Surface plasmons are very sensitive to changes occurring in the immediate surroundings of the metal nanoparticles, including molecular binding events on the sensor surface. The presence of biomolecules on surface alters the local refractive index causing a shift of the plasmon towards higher wavelengths [6, 7]. Quantitative characterization of the molecular layer on the surface can be obtained from the location of the peak position in the spectrum. The peak wavelength of the plasmon depends on the metal, on the geometry and on the properties of the surrounding dielectric [8]. The sensitivity of a sensor is proportional to the sharpness of its plasmon, which can be narrowed by enhancing the regularity of the nanostructures and of the patterns [9]. In T-LSPR, the plasmon phenomenon can be observed as an extinction peak in the transmitted light through the metal pattern. The transmission approach is suitable to be scaled to a compact setup and to perform high-density array measurements. In the present work, a compact fully-portable detection system based on T-LSPR is presented and its performance in terms of real-time monitoring of binding events is reported. We employed gold NIs [10] functionalized with single-stranded DNA in order to perform the real-time observation of DNA hybridization, and we compared our T-LSPR setup with a high-end commercial SPR system.

EXPERIMENTAL DETAILS

Fabrication of Nanoislands sensors

NIs were fabricated on Fluorine-doped Tin Oxide (FTO) covered glass by direct thermal evaporation and subsequent thermal annealing. Surface preparation consists in 20 minutes sonication in a 1:1 solution of 2-propanol and acetone, rinsing with ultra pure water and drying with pure nitrogen. A layer of 5 nm of gold was deposited on surface at RT at 3×10^{-6} mbar at a rate of 0.0016 nm/sec. Samples were placed in the oven at 200°C overnight. Figure 1 shows the formation of NIs after the evaporation and annealing processes. Stability of gold nanostructures on surface is achieved by the employment of the FTO covered glass rather than plain glass, in fact, the porous structure of FTO allows the penetration of gold during the annealing process [11]. As a result, NIs are stable in wet environment without any further surface treatment needed.

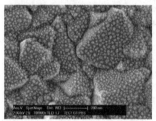

Figure 1. SEM image of gold NIs on FTO covered glass after thermal evaporation and subsequent thermal annealing.

Surface modification of NIs and DNA hybridization

The immobilization of DNA capture single-strands on gold was achieved by sulfur-gold bonds. The NIs FTO-glass slides were incubated at room temperature for 16 hours in Running Buffer (RB; 0.5 M NaCl, 50 mM Na_2HPO_4, pH 6.8) containing the single-stranded DNA (ssDNA) probe. After the incubation, the samples were rinsed with PBS buffer and ultra pure water, and then gently dried with pure nitrogen.
Once the microfluidic was mounted onto the functionalized NIs FTO-glass slides, DNA injections were performed in RB at a flow of 2.5 µl/min for the indicated time. A three-step DNA hybridization protocol was used, which is composed by the subsequent injections of the ssDNA molecules OligoA, OligoB and OligoC.
The sequences of ssDNA used in this work are:
Probe: 5'-CGTACATCTTCTTCCTTTTT-3'-SH (Molecular Weight, MW: 6,132 Da);

OligoA: 5'-AGGAAGAAGATGTACGACCAGCTCAACGAGAAGGTCGCAG-3' (MW
12,451 Da);
OligoB: 5'-TCAACGAGAAGGTCGCAGTAAGTCCTGCGACCTTCTCGTTGAGCTGGT-3'
(MW 14,804 Da);
OligoC: 5'-GACTTACTGCGACCTTCTCGTTGAACCAGCTCAACGAGAAGGTCGCAG-3'
(MW 14,742 Da).
All the oligonucleotide sequences have been purchased from Sigma-Aldrich, Switzerland.

Microfluidics

Microfluidic channels for the delivery of analytes in liquid samples to the surface are built
by mounting a cover plate on top of the NIs FTO-glass slides. Adhesion is obtained thanks to a
black double coated tape (High Performance Double Coated Tape 9086, 3M) that proved to
adhere on the gold NIs and to be resistant to the injection of aqueous solution. The black color
tape screens the contribution of the sensor areas external to the channels. The channels have been
patterned by removing material from the tape by laser micromachining; they feature dimensions
of 0.8 mm × 8 mm × 150 μm, where the height is defined by the tape thickness.
The cover plate is made of Polymethyl Methacrylate (PMMA), a transparent thermoplastic
material frequently used as a substitute of glass. This choice allows us to keep very high
transmission properties and high flexibility in design. Inlets and outlets feature a diameter of
0.68 mm, allowing the usage of standard syringe needles. The PMMA cover plate has been
designed with two identical channels in order to be able to perform parallel measurements and
background noise subtraction. The channels from the NIs sample and from the reference sample
are connected in series, as shown in Figure 2, so that the contribution of the bulk solution is
canceled out by the differential measurement.

Figure 2. Differential microfluidic system showing the NIs sample (top) and the reference
sample (bottom) connected in series.

Transmittance setup

A compact and fully-portable system based on T-LSPR employing electronically-driven
power Light Emitting Diodes (LEDs) in the visible range and integrated instrumentation
amplifiers is presented. Two parallel photodiodes collect the light passing through the sample and
the reference sample, respectively. Our approach is based on the fact that the information of T-
LSPR is contained in the position of the plasmon peak in the spectrum. An algorithm fits the data
obtained with three LEDs and extracts the peak location. It takes into account the spectral
characteristics of all the components, namely, the fact that the LEDs do not emit at a single
wavelength and that the photodiodes have responses that change over the visible range. A
detailed description of the transmittance setup and the peak extraction algorithm can be found in
[12]. The hereby described system is an improvement of the first prototype that includes the

second photodiode to perform real-time differential measurements. The parallel approach allows for light drift and artifacts compensation and it is the first step towards multiple array detection.

RESULTS AND DISCUSSION

NIs stability in aqueous solution

Extinction spectrum of gold NIs on FTO-covered glass slides has been measured prior and after 24 hours immersion in running buffer. Results in Figure 33 show that there is no appreciable shift in the extinction spectrum, demonstrating that the NIs are stable in static conditions. Spectra have been recorded with a commercial spectrophotometer (Ocean Optics 2000+).

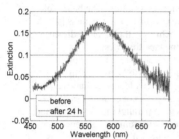

Figure 3. Extinction spectra prior (gray line) and after (black line) 24h NIs immersion in buffer.

NIs adhesion under flow conditions has been tested as well. Microfluidics has been mounted on the NIs sample and the running buffer has been flushed for 10 minutes at 2.5 µl/min. The plasmon peak position has been continuously monitored with our portable setup. The absence of peak-shift proves the stability of gold NIs under flow. In Figure 44 the peak position is represented over time. The standard deviation in the determination of the peak wavelength is 0.2 nm.

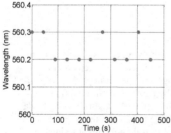

Figure 4. NIs dynamic test: peak position over time recorded for 10 minutes in buffer flow at 2.5 µl/min.

Real-time DNA hybridization

After overnight incubation of gold NIs with Probe DNA, a plasmon peak shift of 6.7 nm has been observed in dry conditions with the spectrophotometer.

Microfluidics has been built on top of ssDNA-functionalized NIs. A three-step DNA hybridization has been performed. At a first step, the OligoA is flushed in the channels. Upon binding with the strand complementary to the Probe DNA immobilized on the surface, OligoA exposes a single-strand trigger complementary to OligoB for the following step (Figure 55.b). The specific binding leads to a 5.7 nm peak shift measured by means of our portable setup. The second and third steps consist respectively in the binding of OligoB on the single-strand trigger of OligoA (Figure 5.c), and in the binding of OligoC on the single-strand trigger of OligoB (Figure 5.d), bringing an overall additional 5.6 nm peak shift.

The same binding experiment has been run on Biacore X100 in order to compare the kinetics obtained with our portable setup. The same reagents and assay conditions have been used on both systems. Results from the two measurement setups are presented in Figure 5.

The dispersed points in step (a) are due to separate injections of OligoB and OligoC in the absence of OligoA, in order to evaluate the selectivity of these sequences towards the Probe DNA on the surface. The subsequent injection of RB drives the baseline back to its original value, proving that there are no molecules bound to the surface. We assume the bulk effect contribution to be more important in our setup, in contrast with the Biacore instrument, giving rise to the signal observed in step (a). Specific binding of DNA sequences is proved in steps (b), (c) and (d), where an increase in resonance units of the Biacore sensorgram and a peak shift in wavelength for our portable setup are shown. The specificity of the complementary DNA strands has been further verified (data not shown).

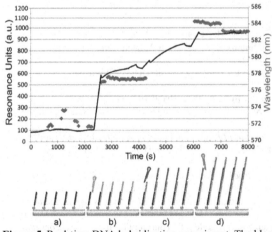

Figure 5. Real-time DNA hybridization experiment. The black line (left axis) represents the SPR measurement from Biacore X100, GE. The gray dots (right axis) represent the peak positions in wavelength as extracted from our setup and algorithm. The following steps are shown: a) buffer baseline with Probe DNA immobilized on the surface. The dispersed points are due to separate

injections of OligoB and OligoC in the absence of OligoA, in order to evaluate the selectivity of these sequences towards the Probe DNA on the surface; b) injection of OligoA; c) injection of OligoB; d) injection of OligoC.

Therefore, we demonstrated here that gold NIs can host heterogeneous DNA hybridization assays. The sensitivity of our sensor can be significantly improved by enhancing the regularity of the nanostructures and surface patterns.

CONCLUSIONS

In this paper we present a fully portable and low-cost detection system based on a differential measurement approach. We developed gold NIs sensors and demonstrate, for the first time, real-time DNA hybridization assays by means of our portable T-LSPR setup. Our setup proves its suitability to observe the real-time binding of molecules on surface. The results, confirmed by the reference experiment performed on a commercially available SPR system, exhibit the same trend for both systems. Implementation of multiple arrays detection together with the further down-scaling of the system, allowed by the usage of low-power components, will make our device fitting the requirements for the diagnostics of the future.

ACKNOWLEDGMENTS

This work, scientifically evaluated by the SNSF, has been funded by Nano-Tera.ch (project 128852 "ISyPeM"), an initiative financed by the Swiss Confederation.

REFERENCES

1. Svedendahl M, Chen S, Dmitriev A, Käll M. *Nano Letters* 9: 4428-33 (2009).
2. Doron-Mor I, Cohen H, Barkay Z, Shanzer A, Vaskevich A, Rubinstein I. *Chemistry – A European Journal* 11: 5555-62 (2005).
3. Ruach-Nir I, Bendikov TA, Doron-Mor I, Barkay Z, Vaskevich A, Rubinstein I. *J Am Chem Soc* 129: 84-92 (2006).
4. Hutter E, Pileni M-P. *The Journal of Physical Chemistry B* 107: 6497-9 (2003).
5. Anker JN, Hall WP, Lyandres O, Shah NC, Zhao J, Van Duyne RP. *Nat Mater* 7: 442-53 (2008).
6. Miller MM, Lazarides AA. *The Journal of Physical Chemistry B* 109: 21556-65 (2005).
7. Malinsky MD, Kelly KL, Schatz GC, Van Duyne RP. *J Am Chem Soc* 123: 1471-82 (2001).
8. Haynes CL, Van Duyne RP. *The Journal of Physical Chemistry B* 105: 5599-611 (2001).
9. Willets KA, Van Duyne RP. *Annu Rev Phys Chem* 58: 267-97 (2007).
10. Cantale V, Simeone FC, Gambari R, Rampi MA. *Sensors and Actuators B: Chemical* 152: 206-13 (2011).
11. Montmeat P, Marchand J-C, Lalauze R, Viricelle J-P, Tournier G, Pijolat C. *Sensors and Actuators B: Chemical* 95: 83-9 (2003).
12. Cappi G, Accastelli E, Cantale V, Rampi MA, Benini L, Guiducci C. *Sens. Actuators, B* In press.

Mater. Res. Soc. Symp. Proc. Vol. 1479 © 2012 Materials Research Society
DOI: 10.1557/opl.2012.1594

Preparation and Characterization of 1-3 BaTiO₃-PVDF Hybrid Nanocomposites

A.A. Rodríguez-Rodríguez, N.A. Morales-Carrillo, C. Gallardo-Vega, G.F. Hurtado-López, J.A. Cepeda-Garza, V. Corral-Flores*
Centro de Investigación en Química Aplicada.
Enrique Reyna Hermosillo 140, 25294, Saltillo, Coah., México.
*Contact author's e-mail: vcorral@ciqa.mx

ABSTRACT

1-3 BaTiO₃-PVDF hybrid nanocomposites were prepared by combining electrospinning, sol-gel and spin-coating techniques. First, one-dimensional structures of barium titanate (BaTiO₃) were obtained by electrospinning. An alcoholic solution consisting of Ba^{2+} and Ti^{4+} ions (1:1 molar ratio) and poly(vinylpyrrolidone) was electrospun at 15 kV, with a tip-to-collector distance of 15 cm and a feed rate of 0.5 mL/h. Ceramic fibers were obtained after sintering the as-spun fibers at 900 °C for 2 hours. In a second step, poly(vinylidene fluoride) (PVDF) was incorporated to the oxide fibers by spin-coating a dimetilformamide solution, thus conforming 1-3 ceramic-polymeric hybrid nanocomposites on stainless steel substrates.

Scanning electron microscopy images showed that the as-spun fibers were smooth, long and continuous with an average diameter of 155 ± 40 nm, ranging from 60 to 240 nm, while sintered fibers presented a reduction in size, with an average diameter of 115 ± 16 nm, ranging from 96 to 120 nm. Sintered nanofibers were also long and continuous but with a rough surface. X-ray diffraction confirmed the perovskite-type structure of the BaTiO₃. A structure refinement revealed a degree of tetragonality of 1.0046.

The polymer crystalline phases were identified by infrared spectroscopy on ATR mode. This study showed the presence of both β and γ polar phases, and absence of non-polar α phase, according to the characteristic bands for such crystalline phases.

The nanocomposites exhibited a ferroelectric behavior and electrical polarization according to their ceramic and polymeric components.

Keywords: ferroelectric, fiber, nanostructure.

INTRODUCTION

In the last decades, barium titanate –BaTiO₃–has become one of the most important ceramics in the electronic industry, with a wide number of applications on components of technological interest, such: sensors, capacitors, electro-optic devices and transducers. BaTiO₃ presents four crystalline structures: cubic, tetragonal, orthorhombic and rhombohedral. The tetragonal array shows good ferro- and piezoelectric properties [1]. On the other hand, poly(vinylidene fluoride) –PVDF– has a large number of mechanical and electrical properties, which highlights its ferro- and piezoelectricity (the largest among synthetic polymers). Four PVDF polymorphs are known: α, β, γ, and δ. The crystalline phase of interest in ferro- and piezoelectricity is the polar β phase, whose conformation is planar zigzag trans-trans (TT) [2].

The adequate combination of these materials results in the formation of composites with enhanced properties compared with the starting materials that can be designed and tailored to meet special requirements of a particular application, such as energy harvesters when taking advantage of its piezoelectric properties.

EXPERIMENT

Barium acetate (from Alfa Aesar) and titanium (IV) butoxide were used as starting materials (1:1 molar ratio). Ba^{2+} salt was dissolved in glacial acetic acid while Ti^{4+} alkoxide was slowly added to a volume of 2-metoxyethanol anhydrous. These two solutions were mixed with poly(vinylpyrrolidone) (PVP, Mw: 1,300,000) at 5 wt.% to obtain a viscous solution, which was loaded into a plastic syringe with a stainless steel needle of 0.5 mm diameter. Electrospinning was carried out with a tip-to-colector distance of 15 cm, an applied voltage of 15 kV and a feed rate of 0.5 mL/h. Fibers were collected on aluminum foil for 1 hour and sintered at 900 °C for 2 hours in air atmosphere to obtain the ceramic phase. Sintered fibers were carefully transferred to a stainless steel substrate and coated with a 14% PVDF (from Arkema) - dimethylformamide (DMF) solution using spin-coating at a spin rate of 3000 rpm for 30 seconds. The samples were then heat treated at 60 °C for 1 hour in order to promote the crystallization of the β phase on the polymer. All chemicals were purchased from Sigma-Aldrich, otherwise specified, and were used without further purification.

The morphology of the fibers and nanocomposites was observed by scanning electron microscopy (SEM) using a JEOL JSM-7401F. The samples were previously coated with an Au-Pd alloy in a Denton Vacumm Desk II evaporator. The crystalline phase of the $BaTiO_3$ nanofibers was determined by X-ray diffraction with a Siemens D-5000 difractometer with $CuK\alpha$ radiation operated at 25 mA and 35 kV. The samples were scanned in the 2θ range of 20-80°.

The structural conformation of the PVDF in the nanocomposite was studied by infrared spectroscopy. The analysis was done at a Magna IR 550FTIR spectrometer from Nicolet with a μ-ATR PIKE technologies module. Ferroelectric hysteresis loops were measured using a 300 Hz driving signal amplified by a TEGAM HV. Sample response was collected with virtual ground charge/current amplifier. Field and current signals were digitized simultaneously and numerically processed to obtain the electric displacement and polarization. Aluminum contacts were previously deposited over the 1-3 $BaTiO_3$-PVDF hybrid nanocomposites in a TE18P evaporator.

DISCUSSION

SEM micrographs of the as-spun and sintered fibers are shown in fig. 1a) and 1b) respectively. Both types of fibers were randomly oriented, however their morphology is clearly different. As-spun fibers were smooth, large and continuous with an average diameter of 155 ± 40 nm (ranging from 60 to 240 nm) while sintered fibers suffered a reduction in size with an average diameter of 115 ± 16 nm (ranging from 96 to 120 nm) as a result of organic matter loss during heat treatment. Sintered fibers were also long and continuous but with a rough surface, revealing the presence of grains.

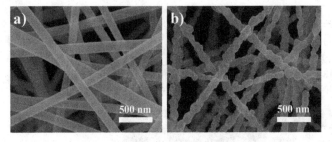

Figure 1. SEM micrographs of the a) as spun fibers and b) sintered fibers.

The X-ray diffraction pattern from the sintered fibers, along with a reference pattern (JCPDS-05-0626), is shown in fig. 2. The diffraction signals correspond to a polycrystalline material with a $BaTiO_3$ perovskite-type structure. A pattern refinement with the software Powder cell 2.4 revealed an average crystallite size (D) of 33.3 nm, considering an equal contribution of Gauss and Lorentz peak shape, and a degree of tetragonality (DoT) of 1.0046. These results are in good agreement with values previously reported [3,4]. Cell parameters are shown in table I.

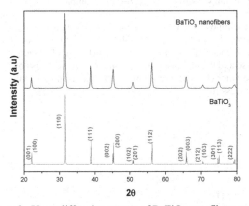

Figure 2. X-ray diffraction pattern of $BaTiO_3$ nanofibers and $BaTiO_3$ reference.

Table I. Cell parameters (a and c), degree of tetragonality (DoT) and crystallite size (D) obtained from the X-ray pattern refinement.

Parameter	Value
a (Å)	4.0172
c (Å)	4.0356

DoT	1.0046
D (nm)	33.3

Fig. 3 shows SEM micrographs of the cross-sectional view of a BaTiO$_3$-PVDF composite, with average thickness of ~18 μm. As shown in fig. 3(b), the nanofibers were completely embedded in the PVDF matrix conforming a hybrid composite with a 1-3 type connectivity.

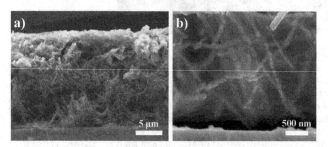

Figure 3. Micrographs of the cross-sectional view of the BaTiO$_3$-PVDF sample.

Fig. 4 shows the infrared spectra of the composite and the as-received PVDF in powder form. The α, β and γ phases are mainly identified by the bands at 762, 840 and 1235 cm^{-1} respectively [5]. The three phases are present in the as received polymer, while the BaTiO$_3$-PVDF sample showed predominantly the polar β phase and the absence of the non-polar α phase. This could be associated to a rearrangement of the polymer chain as a result of the shear and elongational forces produced by the spin-coating process [6].

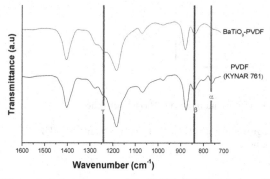

Figure 4. Infrared spectra of the BaTiO$_3$-PVDF sample and as received PVDF.

The beta fraction of the BaTiO$_3$-PVDF sample was calculated by the equation 1 proposed for Gregorio and Cestari [7].

$$F\ (\beta) = A_\beta\ /(1.26\ A_\beta + A_\alpha) \qquad (1)$$

Where A_β and A_α are the absorption band intensity for α and β, respectively. 1.26 is the ratio between the absorption coefficients $K\alpha$ (6.1×10^4 cm^2/mol) and K_β (7.7×10^4 cm^2/mol).

As can be seen in fig. 5 (a), the BaTiO$_3$-PVDF hybrid nanocomposite exhibited a ferroelectric response, with a maximum polarization of 5.97 μC/cm^2 and coercive field of 57.5 kV/cm. As a comparison, a PVDF film was also measured placing an electric contact on a region of the sample where only polymer was present. The PVDF film also presented a ferroelectric behavior, with a maximum polarization of 1.17 μC/cm^2 and coercive field of 48.7 kV/cm. The difference in the values of polarization and coercive field is due to the presence of the ceramic phase. The overall ferroelectric response in the composite suggests a good coupling between the ceramic and polymeric components.

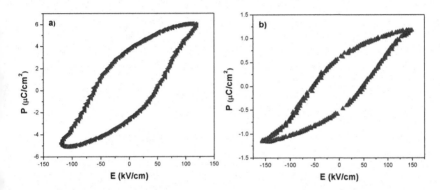

Figure 5. Hysteresis loops of a) the hybrid BaTiO$_3$-PVDF nanocomposite and b) the PVDF film.

CONCLUSIONS

1-3 BaTiO$_3$-PVDF hybrid nanocomposites were successfully obtained. Electrospinning and spin-coating were adequate techniques for the preparation of this kind of composite under the processing conditions used in this work. XRD and FTIR confirmed the presence of the polar phases in the ceramic fibers and the polymer, respectively. Both the hybrid nanocomposite and the PVDF film exhibited ferroelectric hysteresis.

ACKNOWLEDGMENTS

This research was funded by Conacyt, under the project number 133815.

REFERENCES

1. T. K. Kundu, A. Jana & P. Barik, Bull. Mater. Sci. 31, 501-502 (2008).

2. K. Khalil, Mat. Res. Innovat 2, 256-262 (1999).
3. H. Li, H. Wu, D. Lin & W. Pan, J. Am. Ceram. Soc. 92, 2162-2164 (2009).
4. M.B. Smith, K. Page, T. Siegrist, P.L. Redmond, E.C. Walter, R. Seshadri, L.E. Brus, and M.L. Steigerwald, J. AM. Chem. Soc. 130, 6955-6963 (2008).
5. A.B. Silva, C. Wisniewski, J.V.A. Esteves, R.G. Jr., J. Mater. Sci. 45, 4206-4215 (2010).
6. S. Ramasundaram, S. Yoon, K. J. Kim, J.S. Lee, Macromol. Chem. Phys. 200, 2516-2526 (2008).
7. R. Gregorio, M. Cestari, J. Polym. Sci. Part B: Polym. Phys. 32, 859 (2004).

Mater. Res. Soc. Symp. Proc. Vol. 1479 © 2012 Materials Research Society
DOI: 10.1557/opl.2012.1595

Effect of Ligands on the Dispersion of Ni Nanoparticles in Ni/SBA-15 Hydrogenation Catalysts

Ortega Domínguez R.A., Peñaloza Orta C., Puente Lee I., Salcedo Luna C. and Klimova T.

Facultad de Química, Departamento de Ingeniería Química, Universidad Nacional Autónoma de México (UNAM), Cd. Universitaria, Coyoacán, México D.F., 04510, México.

E-mail: klimova@unam.mx

ABSTRACT

In the present work, a comparison study of the Ni catalysts supported on SBA-15 silica support prepared with and without the addition of organic ligands (citric acid and ethylenediaminetetraacetic acid (EDTA) was undertaken. The aim of this study was to inquire on the effect of the addition of organic ligands on the characteristics of the supported NiO and Ni nanoparticles and on their activity and selectivity in hydrogenation (HYD) of aromatics. Catalysts with different metal loadings (5, 10 and 20 wt. % of Ni) were prepared, characterized by nitrogen physisorption, small-angle and powder XRD, TPR, UV-vis DRS, and HRTEM, and tested in HYD of naphthalene (NP). It was found that Ni(II)-Ligand complexes were formed in aqueous solutions of Ni(NO$_3$)$_2$ containing citric acid (CA) or EDTA. Catalysts prepared from impregnation solutions with and without ligands had different textural characteristics and dispersion of NiO particles after calcination at 500 °C for 4 h. As it was shown by XRD, DRS and TPR, dispersion of NiO particles significantly increased when EDTA was used, whereas it noticeably decreased after the addition of CA. Similar trends were observed in the dispersion of metallic Ni particles after reduction of the NiO/SBA-15 precursors (HRTEM). In line with the characterization results, catalytic activity tests revealed strong differences in the activity of the prepared Ni/SBA-15 catalysts in hydrogenation of naphthalene. Catalysts prepared with the addition of EDTA were more active than those prepared without ligands. On the contrary, the HYD activity of a series of the Ni catalysts prepared with citric acid was lower than of other corresponding samples. The reasons of such a different behavior of the catalysts prepared with two organic ligands used are discussed on the basis of the obtained characterization results.

KEYWORDS: Ni, Nanostructure, Hydrogenation

INTRODUCTION

Nickel catalysts supported on silica, alumina, or silica-alumina materials are used in many industrial catalytic processes such as hydrogenation, hydrogenolysis, reforming, gas-shift conversion, etc. due to its easy availability, high activity and low cost [1-3]. To produce clean fuel from high boiling oil fractions containing large amount of aromatic hydrocarbons hydrogenation catalysts with superior catalytic performance are needed [4]. Hydrogenation of aromatics reduces their content in petroleum derived fuels and improve their quality. Since catalytic reaction takes place on the surface of metal particles, their dispersion, the physical and chemical properties determine the activity and selectivity of the catalyst. The above characteristics depend on a number of factors such as a support nature and its interaction with the deposited metal species, metal loading and the technique used for the catalyst preparation. In the present work, a comparison study of the Ni catalysts supported on SBA-15 silica prepared with and without the addition of organic ligands (citric acid and ethylenediaminetetraacetic acid) was undertaken. The aim of this study was to inquire on the effect of the addition of organic ligands on the characteristics of the supported NiO and Ni nanoparticles and on their activity and selectivity in hydrogenation (HYD) of aromatics.

EXPERIMENTAL DETAILS

Synthesis of SBA-15 was carried out by the method reported by Zhao et al. [5,6]. The triblock copolymer Pluronic P123 was used as the structure-directing agent and tetraethyl orthosilicate as the silica source. Catalysts with different metal loadings (5, 10 and 20 wt. % of Ni) were prepared by incipient wetness impregnation of the SBA-15 support with aqueous solutions of $Ni(NO_3)_2 \cdot 6H_2O$ with citric acid (CA, $C_6H_8O_7 \cdot H_2O$) or ethylenediaminetetraacetic acid (EDTA) or without any ligand. The following molar ratios were used in the catalyst preparation: Ni:CA = 1:2 and Ni:EDTA = 1:1. After impregnation, catalysts were dried (100 °C, 12 h) and calcined in air atmosphere (500 °C, 4 h). As all the catalysts were supported on SBA-15, hereinafter, we will denote them only as XNiL, where X represents nickel loading (wt. % of Ni), and L (L = CA, EDTA or nothing) indicates if any ligand was used in the preparation.

SBA-15 support and Ni catalysts were characterized by nitrogen physisorption, small-angle and powder X-ray diffraction (XRD), temperature-programmed reduction (TPR), UV-vis diffuse reflectance spectroscopy (DRS), SEM-EDX and high resolution transmission electron microscopy (HRTEM). The HYD activity tests were performed in a batch reactor at 300 °C and 7.3 MPa total pressure for 6 h using hexadecane solution of naphthalene (NP) (1 wt. % of NP). Before the activity tests, the catalysts were reduced *ex situ* in a tubular reactor under H_2/Ar (20 % H_2) flow at 400 °C for 4 h. The reduced catalysts (0.1 g) were transferred in an inert atmosphere (Ar) to a batch reactor with 40 ml of NP solution. The course of the reaction was followed by withdrawing aliquots each hour and analyzing them by GC.

RESULTS AND DISCUSSION

Characterization of impregnation solutions and prepared catalysts

UV-vis spectra of the impregnation solutions used for the preparation of the catalysts are shown in figure 1. In absence of ligands, nickel nitrate showed three main absorption bands: two sharp bands at 300 and 393 nm and one broad band in the 550-800 nm region. All these bands are characteristic for octahedrally coordinated hydrated Ni^{2+} ions [7,8]. When EDTA was added to the impregnation solution containing Ni^{2+} species, remarcable changes were observed in the UV-vis spectrum (figure 1,c). Thus, the sharp band shifted from 393 to 377 nm, and the broad band transformed into a symmetrical band with a maximum at 585 nm. A blue shift of the above electronic transitions is consistent with the formation of a Ni(II)-EDTA complex. In the case of citric acid, changes in the UV-vis spectrum of Ni^{2+} were less strong than those observed for EDTA. A broad band observed in the 500-800 nm region was less symmetrical (figure 1,b), and it had a maximum at 635 nm and a shoulder near 740 nm. This is in line with the fact that EDTA is a stronger ligand than citric acid.

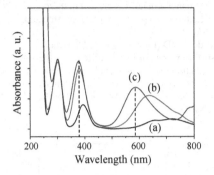 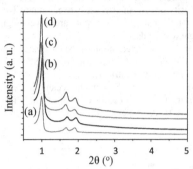

Figure 1. UV-vis spectra of solutions used for catalyst preparation: (a) $Ni(NO_3)_2$, (b) $Ni(NO_3)_2$ + CA, (c) $Ni(NO_3)_2$ + EDTA.

Figure 2. Small-angle XRD patterns of (a) SBA-15 support and corresponding catalysts: (b) 10Ni, (c) 10NiCA, (d) 10NiEDTA.

Small-angle XRD patterns of SBA-15 support and calcined XNiL/SBA-15 catalysts (figure 2) showed the reflections characteristic for *p6mm* hexagonal arrangement of mesopore structure of the SBA-15 material. This result confirmed that the characteristic ordered pore structure of the SBA-15 support was preserved in all catalysts. However, catalysts prepared from impregnation solutions with and without ligands had different textural characteristics after calcination at 500 °C for 4 h (table I).

Table I. Textural characteristics[a] of SBA-15 support and calcined XNiL catalysts, and size of supported NiO nanoparticles[b].

Sample	S_{BET} (m²/g)	S_μ (m²/g)	V_P (cm³/g)	D_P (Å)	NiO nanoparticles size (nm)
SBA-15	926	153	1.10	77	
10Ni	733	156	0.84	70	17
10NiCA	432	64	0.82	78	39
10NiEDTA	529	63	0.89	71	-
20Ni	565	195	0.61	70	18
20NiCA	360	59	0.71	74	48
20NiEDTA	410	41	0.66	63	~ 6

[a] S_{BET}, specific surface area calculated by the BET method; S_μ, micropore area determined by the t-plot method; V_P, total pore volume; D_P, pore diameter corresponding to the maximum of the pore size distribution obtained from the adsorption isotherm by the BJH method.
[b] Calculated from powder XRD data using Scherrer equation.

In general, BET surface area and total pore volume decreased with the increase in nickel content in the samples. However, Ni catalysts prepared with the addition of citric acid showed lower specific surface area than other catalysts with the same metal loading. In addition, a significant increase in the micropore area was observed in the Ni/SBA-15 samples prepared without ligands, being this effect more notorious with Ni loading. Probably, some secondary microporosity due to small voids between deposited NiO nanoparticles was created in these samples. On the contrary, micropore areas were significantly decreased in the catalysts prepared using organic ligands.

Further characterization of the catalysts by powder XRD (figure 3) showed that the dispersion of NiO species supported on the SBA-15 surface was significantly different in the samples prepared with citric acid, EDTA and without addition of organic ligands. Unexpectedly, the use of citric acid resulted in a noticeable increase in the size of NiO nanoparticles, whereas the effect of EDTA was contrary (figure 3, table I). Good dispersion of nickel species was found in the XNiEDTA samples for Ni loadings up to 20 wt. %.

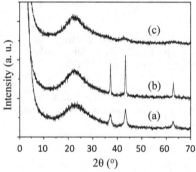

Figure 3. XRD patterns of SBA-15-supported catalysts: (a) 10Ni; (b) 10NiCA and (c) 10NiEDTA.

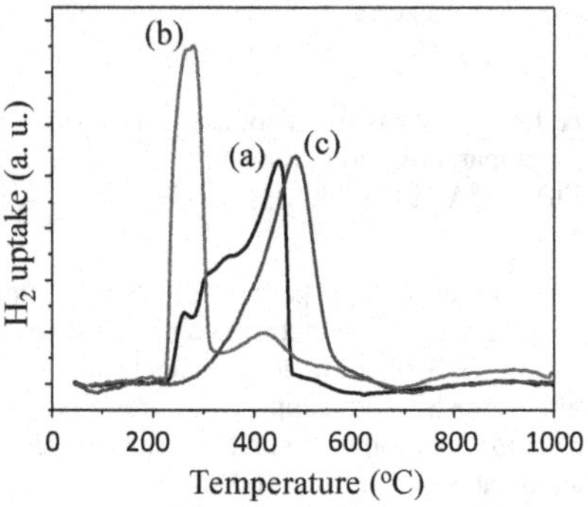

Figure 4. TPR profiles of SBA-15-supported catalysts: (a) 10Ni; (b) 10NiCA and (c) 10NiEDTA.

TPR characterization results (figure 4) showed that the addition of ligands in the impregnation solutions also changed the reduction behavior of NiO species. In the TPR profile of the sample prepared without ligands, hydrogen consumption in a broad temperature interval (from 220 to 480 °C) was observed. This evidenced the presence of a mixture of NiO particles with different characteristics. According to literature [4], the hydrogen consumption at low temperature interval (up to 400 °C) can be ascribed to the reduction of "free" crystalline nickel oxide and the dispersed NiO phase deposited on silica surface, whereas reduction signals observed at higher temperatures are due to NiO species is some interaction with the support. Addition of ligands (CA and EDTA) resulted in a less number and better definition of the Ni^{2+} reduction signals. In the case of EDTA, onle one reduction peak with the maximum at 480 °C was observed indicating the presence of only dispersed NiO particles in a stronger interaction with the SBA-15 support. On the contrary, the use of citric acid resulted in appearance of two signals: i) a sharp and intence peak located between 200 and 300 °C, which can be attributed to the reduction of large NiO crystals, and ii) a lower intensity signal at 420 °C attributable to dispersed NiO particles in weak interaction with the support.

Dispersion of metallic Ni nanoparticles in different series of the prepared samples also followed the trends similar to those described above. XRD and HRTEM characterization of the catalysts after reduction at 400 °C for 4 h showed that the dispersion of Ni increased in the following order: XNiCA < XNi < XNiEDTA. Thus, only small Ni nanoparticles (less than 5-6 nm diameter) were observed on the HRTEM micrograph of the reduced 20NiEDTA catalyst, whereas a mixture of small and very large particles (~ 60 nm) were found in the 20NiCA sample (figure 5).

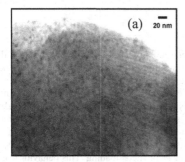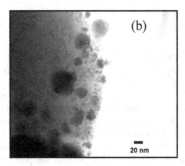

Figure 5. HRTEM micrographs of reduced 20NiEDTA (a) and 20NiCA (b) catalysts.

Catalytic activity

Catalytic activity of the reduced Ni/SBA-15 catalysts was evaluated in hydrogenation (HYD) of naphthalene. Figure 6 shows corresponding general reaction network. It can be seen that first naphthalene is transformed in tetraline, which can suffer further hydrogenation leading to the formation of a mixture of cis- and trans-decalines.

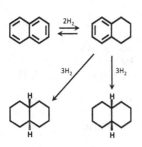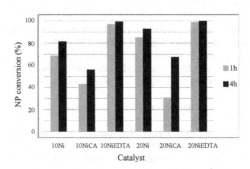

Figure 6. General reaction network for hydrogenation of naphthalene.

Figure 7. Naphthalene conversions obtained with XNiL catalysts at different reaction times.

43

Figure 7 shows naphthalene conversions obtained with different Ni catalysts at 1 and 4 h reaction time. Catalytic activity tests revealed strong differences in the activity of the reduced Ni/SBA-15 catalysts. Catalysts prepared with EDTA showed exceptionally high HYD activity. Naphthalene conversions higher than 90 % were obtained with these catalysts in 1 h (figure 7). On the contrary, catalytic activity of the samples prepared with CA was even smaller than of the corresponding reference Ni catalysts prepared without organic ligands. The above catalytic trends are well in line with the characterization results, evidencing that the addition of EDTA improves the dispersion of oxide and reduced Ni nanoparticles, whereas citric acid leads to the formation of larger NiO and Ni particles than in the similar samples prepared without ligands.

CONCLUSIONS

In the present work, a series of Ni catalysts supported on SBA-15 was prepared with and without addition of organic ligands to the impregnation solutions. It was observed that the use of organic ligands is an easy way which allows modifying dispersion of nickel nanoparticles obtained on the SBA-15 surface after calcination and reduction. However, results obtained with two ligands tested, both able to form complexes with Ni^{2+} in aqueous solutions, were completely different. Thus, addition of EDTA to the impregnation solutions used for catalyst preparation had a beneficial effect on the dispersion of Ni species and resulted in the Ni/SBA-15 catalysts with high hydrogenation activity. On the contrary, the activity of a series of the Ni catalysts prepared with citric acid was lower than of other samples with the same metal loading. This behavior was attributed to larger size of NiO and Ni nanoparticles in the samples prepared with CA.

ACKNOWLEDGMENTS

Financial support by CONACYT, Mexico (grant 100945) is gratefully acknowledged. The authors thank M. Aguilar Franco for technical assistance with small-angle XRD.

REFERENCES

1. M.P. González-Marcos, J.I. Gutiérrez-Ortiz, C. González-Ortiz de Elguea and J.R. González-Velasco, *J. Molec. Catal. A: Chem.* **120**, 185 (1997).
2. K.D. Ghuge, A.N. Bhat and G.P. Babu, *Appl. Catal. A: Gen.* **103**, 183 (1993).
3. F.-W. Chang, M.-T. Tsay and M.-S. Kuo, *Thermochim. Acta* **386**, 161 (2002).
4. J.R. Grzechowiak, I. Szyszka, J. Rynkowski and D. Rajski, *Appl. Catal. A: Gen.* **247**, 193 (2003).
5. D. Zhao, Q. Huo, J. Feng, B.F. Chmelka and G.D. Stucky, *J. Am. Chem. Soc.* **120**, 6024 (1998).
6. D. Zhao, J. Feng, Q. Huo, N. Melosh, G.H. Frederickson, B.F. Chmelka and G.D. Stucky, *Science* **279**, 548 (1998).
7. I. Bonneviot, O. Legendre, M. Kermarec, D. Olivierand M. Che, *J. Colloid. Interface Sci.* **134**, 534 (1990).
8. K.-Q. Sun, E. Marceau and M. Che, *Phys. Chem. Chem. Phys.* **8**, 1731 (2006).

Mater. Res. Soc. Symp. Proc. Vol. 1479 © 2012 Materials Research Society
DOI: 10.1557/opl.2012.1596

Porphyrin Dendrimers

Flores-Rojas Guadalupe G.[1], Martínez-Klimov Mark E., Morales-Saavedra Omar G.[2],
and Martínez-García Marcos [1,*]

[1]Instituto de Química, Universidad Nacional Autónoma de México, Cd. Universitaria,
Circuito Exterior, Coyoacán, México D.F. C.P. 04510, Mexico.
[2]Centro de Ciencias Aplicadas y Desarrollo Tecnológico, Universidad Nacional
Autónoma de México, CCADET-UNAM, Circuito exterior S/N, Ciudad
Universitaria, México D.F. C.P. 04510, México.

ABSTRACT

Dendrons with a porphyrin core and π-conjugated dendron branches have been
synthesized and characterized. The dendrons showed an all *trans* configuration. Cubic
non-linear optical behavior of the styryl and porphyrin-containing dendrimers was
tested *via* Z-Scan measurements in spin-coated film samples.

Keywords: Porphyrin, Dendrimers, NLO-properties

INTRODUCTION

The design and synthesis of π-conjugated dendrimers has been explored extensively
as active chemical components in a wide range of electronic and optoelectronic
devices[1]. Dendrimers have attracted much attention because of their excellent
electroluminescent and electroconductive properties, the globular shape of dendrimers
provides a large surface area that can be decorated with different chromophore
species, thus resulting in a large absorption cross-section and enabling efficient
capture of photons[2]. The nonlinear optical (NLO) properties of several dendrimers
have been published[3]. Here, we report the synthesis and cubic NLO behavior of first
and second-generation dendrimers with π-conjugated branches and a porphyrin core.

EXPERIMENT

Materials and equipments

Solvents and reagents were purchased as reagent grade and used without further
purification. Acetone was distilled over calcium chloride. ^{1}H- and ^{13}C-NMR were
recorded on a Varian-Unity-300 MHz with tetramethylsilane (TMS) as an internal
reference. Infrared (IR) spectra were measured on a spectrophotometer Nicolet FT-
SSX. Elemental analysis was determined by Galbraith Laboratories Inc. (Knoxville,
TN, USA). Electrospray mass spectra were taken on a Bruker Daltonic, Esquire 6000.

Synthesis of dendrimers

A mixture of the respective dendron **1** or **2** (1 mmol), potassium carbonate (21.2 mmol), and 18-crown-6 (0.56 g, 2.12 mmol) in dry acetone (80 mL) was heated to reflux and stirred vigorously in nitrogen atmosphere for 20 min. porphyrin **3** (0.125 mmol) dissolved in dry acetone (40 mL) were added dropwise and the reaction was continued for 7 days. The mixture was cooled and the precipitate was filtered. The filtrate was evaporated to dryness and the dendrimers were purified.

Dendrimer 4. Yield 0.66 g (55 %), purple solid, mp 226 °C, UV–vis CH_2Cl_2 (nm): 231, 302, 422, 454, 518, 556, 595, 651, 694. IR (KBr, cm^{-1}): 3434, 2056, 3026, 2922, 2857, 1655, 1636, 1600, 1500, 1453, 1406, 1379, 1284, 1235, 1174, 1111, 1068, 1015, 987, 960. ^{1}H NMR (300 MHz, CDCl$_3$) δ_H (ppm): -2.68 (s, 2H), 5.3 (s, 8H), 7.25 (s, 40H), 7.28 (d, 2H, J=1.5 Hz), 7.34 (d, 8H, J= 9 Hz), 7.35 (s, 4H,), 7.38 (d, 8H, J= 14.7 Hz), 7.41 (d, 8H, J= 14.7 Hz), 7.43 (t, 2H, J= 1.5 Hz), 7.58-7.61 (m, 4H,), 8.11 (d, 8H, J= 9.0 Hz), 8.87 (s, 8H,). ^{13}C NMR (75 MHz, CDCl$_3$) δ_C (ppm): 70.3, 113.1, 119.7, 124.6, 125.4, 126.6, 127.8, 128.2, 128.7, 129.6, 135.0, 135.6, 137.2, 137.9, 138.2, 158.6. ESI MS (m/z): 1856. Anal. Calcd. for $C_{136}H_{102}N_4O_4$: C, 88.09, H, 5.53, N, 2.95 %, Found: C, 88.06, H, 5.55 %.

Dendrimer 5. Yield 0.237 g (52 %), purple solid, mp 294 °C, UV–vis CH_2Cl_2 (nm): 231, 302, 422, 454, 518, 556, 595, 651, 694. IR (KBr, cm^{-1}): 3434, 2056, 3026, 2922, 2857, 1655, 1636, 1600, 1500, 1453, 1406, 1379, 1284, 1235, 1174, 1111, 1068, 1015, 987, 960. ^{1}H NMR (300 MHz, CDCl$_3$) δ_H (ppm): -2.68 (s, 2H), 5.32 (s, 8H), 7.26 (s, 80H), 7.29 (t, 6H, J=1.5 Hz), 7.36 (d, 8H, J= 9 Hz), 7.39 (d, 24H, J= 14.7 Hz), 7.43 (d, 24H, J= 14.7 Hz), 7.45 (t, 6H, J= 1.5 Hz), 7.58-7.61 (m, 24H), 8.12 (d, 8H, J= 9.0 Hz), 8.88 (s, 8H). ^{13}C NMR (75 MHz, CDCl$_3$) δ_C (ppm): 71.3, 113.2, 119.7, 124.4, 125.4, 126.1, 127.8, 128.3, 128.7, 129.6, 135.1, 134.9, 136.9, 137.5, 138.2, 158.9 (. ESI MS (m/z): 3510. Anal. Calcd for $C_{264}H_{198}N_4NaO_4$: C, 90.25, H, 5.68, N, 1.59 %, Found: C, 90.22, H, 5.64%.

DISCUSSION

Dendrons of first and second generations containing styryl groups were prepared using the Heck reaction in agreement with the literature data[4] (Scheme 1).

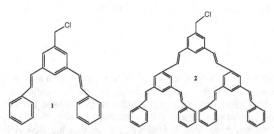

Scheme 1. dendrons of first **1** and second **2** generations

The synthesis of dendrimers **4** and **5** involves *O*-alkylation of dendrons **1** or **2** with porphyrin **3** (Scheme 2). The reaction was carried out in acetone and K_2CO_3 at reflux for 7 days.

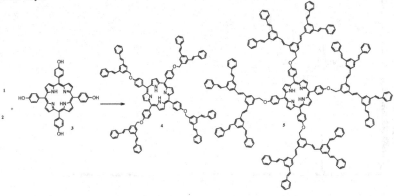

Scheme 2. Synthesis of OPV-phorphyrin dendrimers

The ^{1}H NMR spectra of the dendrimers **4** and **5** (Figures 1 and 2) showed one broad signal at δ_H -2.68 due the protons inside the porphyrin ring, a broad signal at δ_H 5.3, and 5.32 due to the-CH_2-O protons. The vinylic protons give two doublets, at δ_H 7.38 and 7.41 both with a couplet constant J= 14.7 Hz. In both cases a broad peak is also observed at δ_H 7.25-7.43 assigned to the aromatic protons. Finally, one singlet was observed at δ_H 8.87 due to the protons at the pyrrole ring.

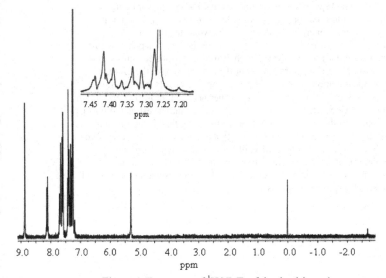

Figure 1. Espectrum of ^{1}H NMR of the dendrimer **4**

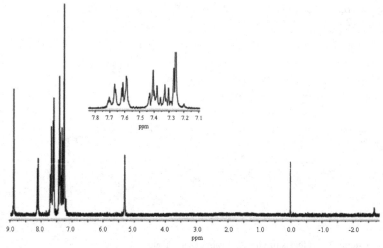

Figure 2. Espectrum of ^{1}H NMR of the dendrimer **5**

Linear and third order nonlinear optical properties

The dendrimers **4** and **5** were spin-coated onto glass substrates (at ~0.33 g L^{-1} concentration, Mw: 1854.76 g mol^{-1} and 3487.54 g mol-1, with average film thickness of ~100 nm and were used for cubic NLO-characterization via the Z-Scan technique. Figure 3a shows the optical spectra in the visible range 400-600 nm, for both dendrimers indicating additional conjugation of delocalized π-electrons and multi-directional charge transfer properties. Under this framework, the available laser excitation line for Z-Scan experiments (@ λ_{Z-Scan}=632.8 nm) is also depicted in this figure 3 (vertical dashed line). At this wavelength, lower absorptive conditions occur, allowing non-resonant NLO-characterizations for these samples and, therefore, avoiding fast photo-degradation, thermal and self-absorptive effects during long Z-Scan measurements as much as possible. The α_0 values are used for the determination of the nonlinear refractive coefficients. The film thicknesses and the estimated linear refractive indices are also shown in Table 1. It is worth noting that the absorption coefficient of the 4-based film is much larger than that obtained for the 5-based film, in agreement with the molecular weight (Mw) of compound **4** is, proportionally, smaller and the film samples were prepared from similar starting solutions at ~0.33 g L^{-1} concentration. Indeed, the larger amount of molecules within the 4-based film allows stronger absorptive conditions.

Table 1. Linear and nonlinear optical coefficients of the compounds **4** and **5**.

Dendrimer Film Sample	Linear Refractive Index: n_0 (@ 632.8 nm)	Linear Absorption Coefficient: α_0 (@ 632.8 nm) [m^{-1}]	Film Thickness [nm]	NLO-Refractive Index: γ / n_2 Z-Scan @ λ = 632.8 nm $\times 10^{-8}$ [m^2 W^{-1}] / $\times 10^{-1}$ [esu]
Compound 4	1.70 ± 0.05	2,307,841	90	-1.8 / -0.7
Compound 5	1.74 ± 0.05	1,050,898	115	+42.6 / +1.6

NLO/-Scan measurements were performed at room temperature on the deposited dendrimer films. The observed non-local effect of these samples is shown in Figure 3b-c. Theoretical fitting was performed to the experimental data in order to evaluate the nonlinear refractive properties of these compounds. The NLO-response of the deposited films was characterized by varying the polarization input planes of the He-Ne laser system in order to explore microscopic material asymmetries or anisotropies throughout the structure. In general, all NLO-measurements were performed with different laser input polarization states (from 0 to 90°: *S*- to *P*-polarization states, respectively) and the obtained curves are quite similar in each sample, the film structures do not seem to show any significant anisotropic behaviour. As shown in Figure 3b-c, the Z-Scan experimental data obtained from compounds **4** and **5** exhibit typical peak-to-valley/valley-to-peak Z-Scan transmittance curves. The nonlinear refractive response of these samples can be unambiguously determined from these highly symmetrical Z-Scan transmittance curves. We can be concluded that compound **4** exhibits a negative NLO-refractive coefficient (γ or $n_2 < 0$), whereas compound **5** exhibits a positive NLO-coefficient (γ or $n_2 > 0$). The theoretical fits (TFs) to the obtained Z-Scan transmission data (solid lines) are shown in Figure 3b-c. According to Table 1, the TFs allowed us to evaluate a negative NLO-refractive coefficient in the order of $\gamma = -1.8 \times 10^{-8}$ m^2 W^{-1} (or $n_2 = -0.7 \times 10^{-1}$ esu)[1,2] for the 4-based film sample. In contrast, a positive) NLO-refractive coefficient was obtained for compound **5**: $\gamma = +42.6 \times 10^{-8}$ m^2 W^{-1} (or $n_2 = +1.6 \times 10^{-1}$ esu).

One of the most interesting features of our measurements is the sign reversal of the NLO-response observed from the 4-based film to the **5** sample. Since the amorphous films comprise randomly oriented and symmetrical disk-like dendrimer compounds, this result cannot be attributed to local material anisotropies, as has been proven by means of polarization dependent Z-Scan measurements. The length of the dendron units **1** and **2** respectively plays a crucial role in the NLO behavior. Indeed, due to the improved charge transfer and electron mobility throughout the whole OPV phorphyrin dendrimer molecular structures, an enhanced nonlinearity tends to occur in both dendrimer-based film samples, being this γ/n_2 coefficient much larger (in magnitude) for the film composed of bigger **5** dendrimer compounds despite the fact that the 4-based film contains twice as many smaller molecular units as the 5-based

---

[1] $[\gamma] \equiv$ [m^2 W^{-1}].

[2] NLO refractive index in electrostatic units: $n_2(esu) = \left(\dfrac{cn_0}{40\pi}\right)\gamma$

film, according to their respective molecular weights. Nevertheless, in the case of shorter molecular electronic pathways (compound with a shorter excitation cross-section), free electrons are able to in-phase follow the optical field; giving rise to a characteristic NLO-response of the 4-based film. This agrees well with the exhibition of a smaller (in magnitude) NLO refractive index (see Figure 1b). The dendrimer size increases (5-based film) and due to the complex structure of these compounds, the hyperbranched and extremely large electronic pathways exposed by the second generation dendrons 2, provokes that the free electrons follow, in an out-of-phase oscillating configuration, the excitation optical field; thus higher molecular absorption and consequently a stronger NLO refractive property with sign reversal takes place. The sign reversal of the refractive index can be attributed to the switching between in-phase and out-of-phase oscillations of the π-electron density with the molecular size.

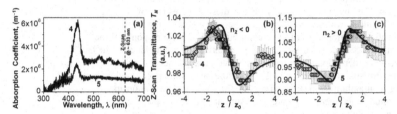

Figure 3a-c. Linear and Z-Scan nonlinear optical measurements for compounds **4** and **5**: Linear absorption coefficients of the dendrimers **4** and **5 (a)**. Closed aperture Z-Scan data obtained for **4 (b)**, and **5 (c)**, based film under similar experimental conditions. Theoretical fitting (TFs): continuous lines.

CONCLUSIONS

The compound **5** exhibits a positive NLO-refractive coefficient in the order of $\gamma = +42.6 \times 10^{-8}$ m^2 W^{-1}. In contrast, compound **4** compound exhibits lower NLO-activity. The observed sign reversal of the NLO-refractive coefficients is attributed to the dendrimer size and to the respective in- and out-of-phase electronic oscillating configurations.

ACKNOWLEDGMENTS

This work was supported by DGAPA-UNAM (IN-202010-3) grant. We would also like to thank Rios O. H., Velasco L., Huerta S. E., Patiño M. M. R., Peña Gonzalez M. A., and Garcia Rios E. for technical assistance.

REFERENCES

1. J. S. Moore, *Acc. Chem. Res.*, **30**, 402 (1997).
2. M. R. Harpham, O. Suzer, C.-Q. Ma, P. Bauerle, T. III. Goodson, *J. Am. Chem. Soc.* **131**, 973 (2009).
3. F. Zeng, S. C. Zimmerman, *Chem. Rev.*, **97**, 1861 (1997).
4. I. V. Lijanova, I. Moggio, E. Arias, R. Vazquez-Garcia, M. Martínez-García, *J. Nanosci. Nanotech.*, **7**, 3607 (2007).

Mater. Res. Soc. Symp. Proc. Vol. 1479 © 2012 Materials Research Society
DOI: 10.1557/opl.2012.1597

Protein Adsorption on Detonation Nanodiamond/Polymer Composite Layers

Lilyana D. Pramatarova[1] , Todor A. Hikov[1], Natalia A. Krasteva[2], Peter Petrik[3], Raina P. Dimitrova[4], Emilia V. Pecheva[1] , Ekaterina I. Radeva[1], Elot Agocs[3], Ivaylo G. Tsvetanov[1] and Radina P. Presker[5]
[1]Institute of Solid State Physics, BAS, Bulgaria,
[2]Institute of Biophysics, BAS, Bulgaria,
[3]Research Institute for Technical Physics and Materials Science, HAS, Hungary,
[4]Institute of Organic Chemistry, BAS, Bulgaria
[5]University of Ljubljana, Slovenia
Corresponding Author: lpramat@issp.bas.bg

ABSTRACT

Composite layers of the detonation nanodiamond/polymer type possess a spatial organization of components with new structural features and physical properties, as well as complex functions due to the strong synergistic effects between the nanoparticles and polymer [1]. Composite layers were deposited by a plasma polymerization (PP) process of the detonation nanodiamond (DND) particles added to a hexamethyl disiloxan (HMDS) monomer [1]. The incorporation of silver ions in the polymer leads to the production of materials that are highly efficient against bacterial colonization and allows better cell adhesion and spreading. [2] For cell culture processes, fibronectin (FN) treatment is one of the commonly used approaches to enhance the cell adhesion on a surface [3].
As an integrated part of our search for improved materials for life science applications such as biomaterials and biosensors, the objective of the present study is to investigate the interaction of Ag-based composite surfaces with FN protein. Two types of composite layers, Ag-ND/PPHMDS and Ag-nano/PPHMDS were obtained by plasma polymerization of HMDS and nanoparticles of Ag and Ag-DND. The composite layers are representative of the different incorporation of the Ag in the polymer net. The structures studied, consisting of composite layers with adsorbed FN were optically characterized with Ellipsometry, Fourier Transform Infrared (FTIR) and Ultra Violet (UV) Spectroscopy as well as by stylus profiling (Talysurf). The kinetic study of the FN adsorption indicates that the process depends on the FN concentration and the exposure time as well as on the surface chemistry of the composites. Compared to the reference sample, all composite layers exhibit an indication of a stronger ability to initiate the intrinsic pathway of coagulation.

Keywords: protein, adsorption, nanostructures

INTRODUCTION

An important goal of materials science is the development of interfaces that integrate the functions of living cells and materials. Nature has given us plenty of ideas on how to build composites and organized structures [4]. The structure of a given biomaterial is crucial when determining the cell response, and respectively, the variants for its biomedical applications. The combined unique properties offered by organic and inorganic constituents within a single material on a nanoscale level make nanocomposites attractive for the next generation of biocompatible materials. In this case, the composite materials of the nanoparticle/polymer type possess a spatial organization of components with new structural features and physical properties, as well as complex functions due to the strong synergistic effects between the nanoparticles and the polymers. Recently, there has been a growing interest in the synthesis of composite functional materials with new physico-chemical properties, involving the integration of inorganic nanomaterials into a polymer matrix [1, 5, 6, 7]. Numerous siloxan-based materials, including polymerized hexamethyldisiloxane (PPHMDS), have been developed. PPHMDS is easy to prepare using a well established procedure of plasma polymerization [2, 8]. PPHMDS has a long history of exploitation in a variety of applications because it is non-toxic, transparent, with a very low surface tension, flexible and it neither dissolves nor swells in a cell culture medium [9]. On the other hand, nanodiamond structures are of interest due to the combination of unique properties inherent to diamond and the specific surface structure of the particles facilitating its functionalization [1].

EXPERIMENT

Materials

The following materials were used in this work:

The substrates used were cover glass (CG) discs with a diameter of 12 mm and silicon (Si). The detonation nanodiamonds (DND), synthesized by a detonation of carbon-containing explosives, are produced with particles with an average size of 4 nm and can be easily modified by appropriate chemical reactions. The diamonds that we used were modified with silver ions, Ag-DND [1]. The incorporation of silver ions in the polymer leads to the production of samples that are highly efficient against bacterial colonization and allows the adhesion and spreading of mammalian cells [2, 10].

Ag-nano particles ("AgBion" from Nnaoindustry, CSSC, Moscow, www.nanotech.ru) are prepared by bio-chemical synthesis in a form of colloidal solution of Ag-nano particles (3-16 nm in size) that are stable in various organic solvents (Dodecane and Octane) as well as in water and a water-ethanol mixture (stabilized by surface-active substances). "AgBion" have anti bacterial, viral and fungus characteristics as well as showing a negative influence to mould and blue-green duckweed.

Polyhexadimethylsiloxan (HMDS) layers were prepared by a plasma polymerization procedure (PPHMDS) as previously described in [8]. The composite deposition was carried out in the same plasma polymerization equipment using the following procedure: the mixture of Ag-DND powder (or AgBion suspension) and HMDS was ultrasonically treated for 15 min. followed by the container with the mixture being stirred (275 r.p.m.) continuously at room temperature. To investigate the influence of the chemistry of the composite surface on protein adsorption one group of the samples was subsequently treated by NH_3 plasma for 5 min [1].

Fibronectin (FN) treatment: the samples (composites) are put in holders, each one of which is situated in a different container which is filled with FN solution. The experiment includes 9 containers with different protein concentrations (1 ug/ml, 10 ug/ml and 20 ug/ml) and different times of immersion for each concentration (1 min, 10 min and 30 min). After the experiment, all samples are repeatedly washed with a phosphate buffered saline solution, PBS and distilled water (3 times each) and put in a sterilie box.

SEM measurements

The surface topography of PPHMDS and composites grown on the Si substrate were examined by SEM (Carl Zeiss NTS GmbH apparatus), using Low Loss BSE Imaging with ZEISS ULTRA GEMINI technology. The thickness of the layers was calculated from the observed cross sections.

Ellipsometry measurements

The ellipsometric measurements were performed by a Woollam M-2000DI rotating compensator ellipsometer in the wavelength range of 193-1690 nm over 706 points at angles of incidence ranging from 55° to 75° by 5°. Most of the ellipsometry results could be fitted using the optical model with the simple Cauchy formula.

FTIR measurements

FTIR spectra of the PPHMDS and nanoparticles/PPHMDS composites were registered by a Bruker FTIR spectrometer at ambient temperature in the range of 400 to 4000 cm^{-1}, using OPUS software, an average of 64 scans and a resolution of 2 cm^{-1}. The assignment of the absorption bands was based on experience with organic compounds and the literature data. The quoted wavelengths are estimated to be within 2 to 3 cm^{-1} of the true values.

Ultraviolet Spectroscopy measurements

UV spectra were recorded on a UV-Visidle spectrophotometer (JASKOV-650 double beam model with single monochromator) in the 200–800 nm wavelength range at room temperature.

RESULTS AND DISCUSSION

In Figure 1, SEM cross-section images of polymer (PPHMDS) (a) and Ag-DND/PPHMDS (b), Ag-nano/PPHMDS (c, d) composites, deposited on Si substrates are shown. For the Ag-DND/PPHMDS (Figure 1-b) two parts of the layer can be distinguished: the dense lower part with a thickness of about 125 nm, and the upper part with a thickness of about 300 nm, a low density (about 0.4 g/cm^3) and a significant concentration of Ag ions. When the polymer is filled with Ag-nanoparticles (Figure 1-c), a significant difference in the layer thickness (about 1000 nm) can be seen compare with the thickness of the PPHMDS (about 300 nm) and Ag-DND/PPHMDS composites. The Ag-nano particles are in fact incorporated in a complex with the polymer and not as lone atoms in its network (Figure 1-d).

After composite deposition, the samples were immersed in the FN solution using the procedure described in the Experimental section. Preliminary investigation included immersion of the CG substrate in the FN solution for 30 mins. at room temperature and their subsequent rinsing with PBS and DI water 3 times.

In Figure 2, FTIR spectra of FN coated CG (line 1) and the same sample rinsed with PBS afterwards (line 2) and 24h later (line 3) are shown. The different character of the spectra point towards an interaction between the CG substrate and the FN due to the experimental conditions.

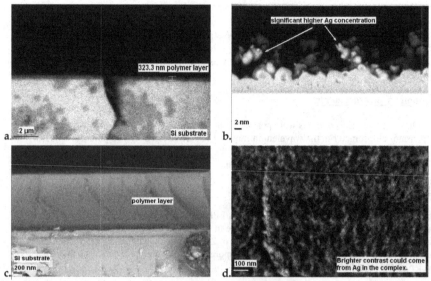

Figure 1 BSE imaging of PPHMDS (a), Ag-DND/PPHMDS (b), Ag-nano/PPHMDS (c, d).

In Figure 3, FTIR spectra of composites (Ag-DND/PPHMDS and Ag-nano/PPHMDS) and their subsequent treatment in NH_3 plasma, followed by immersion in the FN solution can be seen.

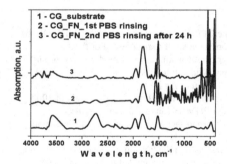

Figure 2 FTIR spectra of FN coated CG (1); the same sample rinsed with PBS afterwards (2); replicated after 24h later (3).

High intensity peaks that are non characteristic of the initial polymer coatings [1] are registered at 1150, 1300 and 1800 cm^{-1} that could be a result of the protein being irreversibly adsorbed on the composites surface. The FTIR results demonstrate that both Ag-DND/PPHDMS (Figure 3-A) and Ag-nano/PPHDMS (Figure 3-B) surfaces significantly interact with the FN. The main difference between the initial layers and those treated with NH_3 plasma layers is in the intensity of the FTIR peaks. It could be concluded that the layer treatment by ammonia plasma modifies its surface chemistry and that the interaction with FN becomes stronger.

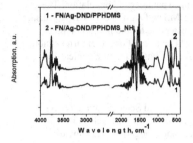 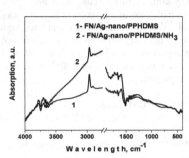

A   B

Figure 3 FTIR (A) spectra of (1) FN/Ag-DND/PPHMDS; (2) treated with NH$_3$ and (B) spectra (1) FN/Ag-nano/PPHMDS; (2) treated with NH$_3$ (B)

Ultraviolet spectra of FN coated composites (Figure 4) show the different state of the silver incorporated in the composite layers which depends largely on the type of the substrate and less on the growth conditions. The layer obtained on CG contains good dispersed silver ions and atoms (peak at 300 nm). On the Si substrate can be seen the presence of silver atom agglomerates (peaks at 240 and 340 nm) as well as nano-sized agglomerates (intense peak at 400 nm).

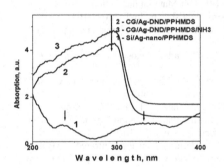

Figure 4 UV spectra of FN coated composite layers: Ag-nano/PPHMDS, grown on Si substrate (1), Ag-DND/PPHMDS, grown on CG substrate (2), and the same composite treated with NH$_3$ plasma (3).

The thickness of the FN/CG/Ag-DND/PPHMDS structure was measured by ellipsometry. The samples were obtained by immersion of the composite in the FN with a concentration 20 μg/ml for 10 min and for 30 min. Thickness was 98.01 nm and 107.40 nm respectively. Due to longer immersion in the FN solution, the protein is adsorbed more.

It could be concluded that FN adsorption depends not only on the protein concentration and exposure time, but also on the surface chemistry of the composites.

CONCLUSIONS

The PPHDMS is a hydrophobic material, which is difficult to wet and used to adsorb proteins or small hydrophobic molecules. Alternatively, its modification with DND and silver cations provide hydrophilic surface coatings. Compared to the reference sample, all the composite layers exhibit an indication of a stronger ability to initiate the intrinsic pathway of coagulation. The FTIR and UV results demonstrated that both Ag-nano/PPHDMS and Ag-DND/PPHDMS surface significantly interact with fibronectin. New peaks, non characteristic for the initial polymer coatings are registered at 1150, 1300 and 1800 cm^{-1} that could be a result of protein irreversible adsorbtion on the surface of the materials. As a result, the surface has different chemical and biological properties.

ACKNOWLEDGMENTS

This work was supported by the Bulgarian Ministry of Education and Science (grant TK-X-1708/2007) the Agency of Innovation of Bulgaria (grant NIF 02-54/2007) as well as by a bilateral project between the Bulgarian Academy of Sciences and Hungarian Academy of Sciences.

REFERENCES

1. L. Pramatarova, E. Radeva, E. Pecheva, T. Hikov, N. Krasteva, R. Dimitrova, D. Mitev, P. Montgomery, R. Sammons and G. Altankov, The advantages of polymer composites with detonation nanodiamond particles for medical applications, On Biomimetincs, ed. by L. Pramatarova, InTech (2011), chapter 14, pp. 297-320.
2. Vasilev K, Sah V, Anselm K, Ndi C, Mateescu M, Dollmann B, Martinek P, Ys, Ploux L & Griesser H J. (2010). Tunable Antibacterial Coatings That Support Mammalian Cell growth. Nano Letter. 10.; 202-207
3. Salmerón-Sánchez M & Altankov G. (2010). Cell-Protein-Material interaction in tissue engineering. Tissue Engineering. Ed. by Daniel Eberli. Published by In-Teh. 077-103
4. Heuer A H et al .(1992). Innovative materials processing strategies: a biomimetic approach. *Science.* 255.; 5048.; 1098 – 1105
5. Shenderova O A, Zhirnov V V & Brenner D W. (2002). Carbon nanostructures. *Solid State Mater.Sci.* 27; (3/4).; 227-356
6. Dolmatov V Yu .(2007). Composite materials based on elastomer and polymer matrix, filled with detonation nanodiamonds. *Russian Chemical Reviews.* 70.; 7.; 607-626
7. Borjanovic V, Lawrence W G, Hens Suzanne, Jaksic M, Zamboni I, Edson C, Vlasov I, Shenderova O & McGuire G E. (2009). Effect of proton irradiation on photoluminescent properties of PDMS-nanodiamond composites. *Nanotechnology.* 19; 1-10
8. Radeva E, Tsankov D, Bobev K & Spassov L. (1993). Fourier Transform Infrared Analysis of Hexamethyldisiloxane Layers Obtained in Low-frequency Glow Discharge. *J. Appl. Polym. Sci.* 50.; 165-171
9. Min-Hsien. (2009). Simple poly(dimethylsiloxane) surface modification to control cell adhesion. *Surf. Interface Anal.;* **41.;** 11–16
10. Agarwal A, Weis T L, Schurr M j, Faith N G, Czuprynski C J, McAnulty J F, Murphy Ch J & Abbott N L. (2009). Surfaces modified with nanometer-thick silver-impregnated polymeric films that kill bacteria but support growth of mammalian cells. *Biomatieials.* dio: 10.1016/j.biomaterials.2009.09092

Mater. Res. Soc. Symp. Proc. Vol. 1479 © 2012 Materials Research Society
DOI: 10.1557/opl.2012.1598

Preparation of Polymer Nanocomposites with Enhanced Antimicrobial Properties

Beatriz L. España-Sánchez[1], Carlos A. Ávila-Orta[1]*, Maria G. Neira-Velázquez[1], Silvia G. Solís-Rosales[1] and Pablo González -Morones[1].

[1]Centro de Investigación en Química Aplicada (CIQA), Blvd. Enrique Reyna H. 140, Saltillo, Coah. C.P. 25294, México. *cavila@ciqa.mx

ABSTRACT

Plasma surface activation and antibacterial properties of nanocomposites of polypropylene/silver nanoparticles (PP/nAg) and nylon-6/silver nanoparticles (Ny6/nAg) were investigated. The nanocomposites were prepared by melt blending assisted by ultrasound, while surface activation was achieved by means of argon plasma. To evaluate the antimicrobial properties of the nanocomposites, pathogen microorganisms such as *Pseudomonas aeruginosa* and *Aspergillus niger* were tested. Scanning Electron Microscopy (SEM) analyses showed a uniform dispersion of nanoparticles within the polymer matrix, though the presence of some agglomerates was also appreciated. On the other hand, surface topography by Atomic Force Microscopy (AFM) suggested that ions from the argon plasma generated ion collisions with the surface of the nanocomposites removing or etching polymer from surface and improving silver nanoparticles exposure, increasing their antimicrobial properties as corroborated by antimicrobial analyses. Nanocomposites exposed to argon plasma presented higher antimicrobial properties than the ones not exposed. These results indicated that plasma treatment increased the contact area of the nanoparticles with the microorganisms and enhanced the antimicrobial properties of nanocomposites. The results also showed that PP/nAg nanocomposites presented higher bacterial inhibition than Ny6/nAg nanocomposites, indicating that the chemical structure of the polymer also plays a big role in the final performance of the composite.

Keywords: Nanocomposites, plasma, antimicrobial.

INTRODUCTION

In recent years, the incorporation of nanometric fillers in polymer matrices has attracted great interest in several fields of science and technology to develop new materials with improved physical, thermal, electrical, and antimicrobial properties, these materials are known as nanocomposites. One of the main problems found in the preparation of polymer nanocomposites, in particular of Polypropylene (PP) and Nylon-6 (Ny6) is the lack of homogeneous dispersion of the nanoparticles into the polymer matrix hindering the enhancement of their bulk properties [1]. In particular, the preparation of nanocomposites with antimicrobial properties results in particles embedded completely in the bulk of the polymer, limiting their interaction with pathogen microorganisms, and reducing severely their antimicrobial behavior [2-5]. Recent researches [3-4, 6-7] indicated that in some cases, a higher concentration of nanoparticles is required in order to observe an inhibitory effect of microorganisms. According to the above, it is also considered that preparation of nanocomposites with enhanced antimicrobial properties is restricted due to the non-uniform dispersion of nanoparticles within the polymer and the low amount of exposed particles having contact with microorganisms such as bacteria and fungus. These limitations can be overcome by using ultrasound assisted melt extrusion to obtain a uniform dispersion of nanoparticles [8], and by surface activation with plasma [9].

In this study, we report the preparation and the surface activation of nanocomposites of Polypropylene/silver nanoparticles (PP/nAg) and Nylon-6/silver nanoparticles (Ny6/nAg). Both were prepared by ultrasound assisted melt blending. The surface activation was carried out with argon plasma aiming to erode the polymer located at the surface and leaving silver nanoparticles more exposed to interact with different pathogens microorganisms.

EXPERIMENTAL

Chemicals. Polypropylene was purchased from Polimeros Mexicanos and has a melt flow index (MFI) of 35g/10 min. Nylon-6 was purchased from Dupont (Zytel) and has a molecular weight (Mw) of 23 000g/mol. Silver nanoparticles were acquired from SkySpring Nanomaterials Inc. with 99.95 % of purity and average size of 20-30 nm.

Bacterial cultures. *Pseudomonas aeruginosa* (ATCC#13388) and *Aspergillus niger* (ATCC#6275). The bacterium was prepared in liquid media using nutritive broth and the fungus was prepared in Potato dextrose agar broth (PDA); both purchased from BD Bioxon.

Nanocomposites preparation. To prepare the nanocomposites the polymers were mixed with the silver nanoparticles using a *Dynisco* extruder with ultrasound system coupled, to prepare the composites, the following conditions were used: 190°C (PP/nAg), 230°C (Ny6/nAg) , 20 kHz and 60 W. Nanocomposites were prepared using 1 and 3% of silver nanoparticles for each polymer. To obtain films of the nanocomposites a Film-maker model Thermospectra was used. The films prepared had diameter of 1 cm and thickness of 50 μm.

Plasma surface activation. Films of the nanocomposites were placed in a cylindrical glass plasma reactor. Plasma was generated using a plasma generator of 13.56 MHz, coupled inductively to the reactor. Films were treated for one hour under argon plasma at 50 W, during the treatment, the pressure inside the reactor was of 1.8×10^{-2} mbar.

Nanocomposites characterization. Nanocomposites films were evaluated by Scanning Electron Microscopy (SEM) in an equipment model Jeol JSM-7041F with a work voltage of 10 kV, samples were fractured with liquid nitrogen. To evaluate the surface topography an Atomic Force Microscopy (AFM) model Veeco Dimension 3000 was used. Thermogravimetric analysis (TGA) was carried out in an equipment Q500 from TA Instruments. Analyses were carried out from 30 to 800 °C at a speed heating of 20 °C/min, under nitrogen atmosphere.

Microbiological analyses. The films were placed in flasks containing nutritive broth and 10^7 FCU/ml of *P. aeruginosa* and left during 24 h, at 37°C. To determinate the inhibition percent in *A. niger*, samples were place in flasks with PDA broth and were incubated during 72 h and 25°C. The antimicrobial activity is determined by comparing the results of the sample containing the antimicrobial agent with a control sample, after a certain contact time.

RESULTS

SEM images of the fractured surfaces of untreated and plasma treated PP/nAg nanocomposites were obtained and the results are presented in Figure 1. Figure 1A represents the surface of untreated PP/nAg 1%, silver nanoparticles are well dispersed in PP, though some nanoparticles clusters are also appreciated, the black arrows in the Figure are indicating their presence. Figures 1B and 1C represent the fractured surfaces of plasma treated PP/nAg 1%. Comparing the SEM images of Figure 1A and Figure 1B, it is appreciated that untreated PP nanocomposites present a smoother surface, whereas plasma treated PP/nAg shows a rougher surface. The surface of the later is rougher since plasma reactive species collide with the nanocomposite surface producing chain scission and remove polymer chains located at the most external part of the nanocomposite, generating roughness of the surface [9-11]. Beake *et al.* [12] reported the surface erosion of PET films treated by plasma. These authors found important changes in the roughness generated by the reaction of argon ions with the polymer surface of PET films, which agrees with the results obtained in this study.

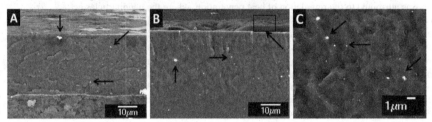

Figure1. SEM images of PP/nAg 1%. A) Untreated, B) Plasma treated (Transversal fracture), C) Plasma treated (Surface).

AFM analysis was carried out in order to visually evaluate the surface erosion generated by plasma treatment. Figure 2 shows topographical AFM images for A) untreated PP/nAg 1% and B) plasma treated PP/nAg 1%. The images were taken in tapping mode. The surface of untreated PP/nAg 1% is very smooth and few dark grey spots, which represent silver nanoparticles are appreciated, on the other hand, the surface of plasma treated PP/nAg 1% (Figure 2B) is rougher and many silver nanoparticles are appreciated. After the plasma etch with argon, the roughness of the surface increased and silver nanoparticles were more exposed to the surface, for this reason in Figure 2B a higher number of silver nanoparticles are appreciated. Figure 2C also represents the topographical AFM image of plasma treated PP/nAg 1%, but this image was obtained at higher magnification in order to appreciate more the topographic features of the plasma treated nanocomposite.

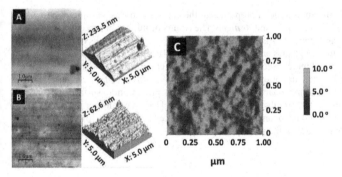

Figure 2. AFM topography images and 3D schemes of A) Untreated PP/nAg 1%, B) Plasma treated PP/nAg 1% and C) Surface topography image of Plasma treated PP/nAg 1%.

The thermal stability of the nanocomposites was studied by thermogravimetric analysis (TGA). TGA curves (residual weight percentage versus temperature) for (A) pristine PP and PP/nAg nanocomposites and for (A) pristine Ny6 and Ny6/nAg nanocomposites are shown in Figure 3. Decomposition temperatures for all polymers and nanocomposites were obtained at 90% (residual weight percentage). In Figure 3A the decomposition temperature of the pristine PP was 402° C, for PP/nAg 1% was 424° C and for PP/nAg 3% was 426° C. Chae *et al.* [15] reported an increase of 14°C in degradation temperature in polypropylene/silver nanocomposites which was attributed to the retardation of heat penetration, where the nanoparticles act preventing the diffusion out of the decomposed polymeric material. In Figure 3B the decomposition temperature of the pristine Ny6 was 408° C, and for Ny6/nAg 1% and Ny6/nAg 3% was 414° C in both cases. For Ny6/nAg nanocomposites, a slight increase in the degradation temperature (6° C) in relation to the pristine polymer was observed, regardless of the silver nanoparticles concentration. Evidently, the decomposition onset for PP/nAg and Ny6/nAg nanocomposites shifts to higher temperature compared to pristine PP and Ny6, the presence of silver nanoparticles improves the thermal stability of the polymers, especially for PP/nAg nanocomposites, where the decomposition temperatures improved in more than 20° C in relation to pristine PP.

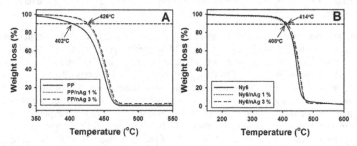

Figure 3. TGA curves of A) PP and PP/nAg nanocomposites and B) Ny6 and Ny6/nAg nanocomposites with 1and 3% of silver nanoparticles.

In order to evaluate the antimicrobial properties of the nanocomposites, antimicrobial tests were carried out with *P. aeruginosa* and *A. niger*. The results of inhibition percentages of *P. aeruginosa* with PP/nAg nanocomposites are presented in Figure 4A, whilst results of inhibition of *A. niger* with PP/nAg nanocomposites are presented in Figure 4B. In both cases, plasma treated nanocomposites resulted more effective in preventing bacteria growth. In some cases these treated nanocomposites were up to four times more effective to kill bacteria than the untreated ones (Figure 4A). Silver nanoparticles in these materials present more surface area to chemically interact with the bacteria inhibiting their development [16]. We consider that antibacterial behavior of untreated nanocomposites can be attributed to the release of silver ions from silver nanoparticles embedded in the polymer matrix [16].

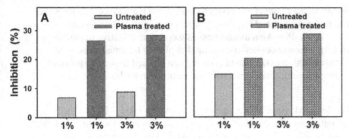

Figure 4. Antimicrobial activity of PP/nAg nanocomposites. Tested with A) *P. aeruginosa*, B) *A. niger*.

Figure 5 shows the antimicrobial effect of Ny6/nAg nanocomposites with *P. aeruginosa* (Figure 5A) and *A. niger* (Figure 5B). Similar to the previous results, in this case the Ny6/nAg nanocomposites that were exposed to plasma showed higher antimicrobial activity than the untreated ones. In both polymers nanocomposites activated with argon plasma increased their antimicrobial activity as is appreciated in Figures 4 and 5. However, the antimicrobial effect was better in PP/nAg than in Ny6/nAg nanocomposites, possibly due to their difference on their chemical structure [17-18]. We suggest that Nylon chemical structure is more stable against plasma attack, thus bond-breaking by plasma reactive species is more difficult requiring more energy to break linkages compared with polypropylene, since the later contains tertiary carbons that are more susceptible to chain scission. Some authors [3-4] have mentioned that the antimicrobial effect of polymer nanocomposites is limited by the direct contact of nanoparticles with the environment, due fundamentally to the immersion of nanoparticles within polymer matrix. Nevertheless, in the case of this study, the plasma treatment plays an important role increasing antimicrobial activity of silver nanocomposites

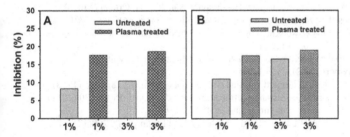

Figure 5. Antimicrobial behavior of Ny6/nAg nanocomposites. Tested with A) *P. aeruginosa*, B) *A. niger*.

CONCLUSIONS

The preparation of polymer nanocomposites with silver nanoparticles by melt blending with ultrasound system is an efficient technique to obtain materials with high degree of dispersion. Surface activation of nanocomposites by means of argon plasma treatment enhances nanocomposite properties due to the removal of polymer chains from the surface, leaving silver nanoparticles more expose to the surface and improving their direct contact with pathogen microorganisms causing cell death. PP/nAg nanocomposites were more effective in inhibiting bacteria growth than Ny6/nAg nanocomposites, indicating that plasma treatment was more effective for the former. Plasma activation or plasma etching can be used to prepare polymer nanocomposites with enhanced antimicrobial properties for their use in health care.

ACKNOWLEDGMENTS

The authors acknowledge Mexican council of science and technology (CONACyT) for their financial support through grant CB2005-49087 and for a scholarship granted to B.L. España (No. 212958). They are also grateful to E. Díaz-Barriga and M. L. López-Quintanilla, for their support in sample preparation for microscopy analyses.

REFERENCES

1. X. Tan, Y. Xu, N. Cai, G. Jia, Polymer Composites, 30(6), 835 (2009).
2. R. Kumar and H. Münstedt, Polymer International, 54(8), 1180 (2005).
3. H. Palza, S. Gutiérrez, K. Delgado, O. Salazar, V. Fuenzalida, J.I. Ávila, G. Figueroa, R. Quijada, Macromolecular Rapid Communications, 31(6), 563 (2010).
4. K. Delgado, R. Quijada, R. Palma, H. Palza, Letters in Applied Microbiology, 53(1), 50 (2011).
5. C. Radheshkumar and H. Münstedt, Reactive and Functional Polymers, 66(7), 780 (2006).
6. S. Egger, R.P. Lehmann, M.J. Heigh, M. Schuppler, Applied and Environmental Microbiology, 75(9), 2973 (2009).
7. A. Kumar, P.K. Vemula, P.M. Ajayan, G. John, Nature Materials, 7(3), 263 (2008).
8. C.A. Ávila Orta, J.G. Martínez Colunga, D. Bueno Báquez, C.E. Raudry López, V.J. Cruz Delgado, P. González Morones, J.A. Valdés Garza, M.E. Esparza Juárez, J.C. Espinosa González, J.A. Rodríguez González, WO/2010/117256.
9. C.A. Ávila Orta, M.G. Neira Velázquez, B.L. España Sánchez, H. Ortega Ortiz, P. González Morones, J.A. Rodríguez González, J.A. Valdés Garza, MX/a/2011/013347.
10. E. Amanatides, D. Mataras, M. Katsikogianni, Y.F. Missirlis, Surface and Coatings Technology, 200(22-23), 6331 (2006).
11. M. Tagawa, K. Yokota, K. Kishida, A. Okamoto, J. Ishizawa, T.K. Milton, High Performance Polymers, 22(2), 213 (2010).
12. B.D. Beake, S.G. Ling, G.J. Leggett, Journal of Materials Chemistry, 8, 1735 (1998).
13. B. Bittmann, F. Haupert, A.K. Schlarb, Ultrasonics Sonochemistry, 18(1), 120 (2011).
14. R.M. France and R.D. short, Langmuir, 14(17), 4827 (1998).
15. D.W. Chae and B.C. Kim, Macromolecular Materials and Engineering, 290(12), 1149 (2005).
16. X. Wu, J. Li, L. Wang, D. Huang, Y. Zuo, Y. Li, Biomedical Materials, 5(4), 044105 (2010).
17. N. Perkas, G. Amirian, S. Dubinsky, S. Gazit, A. Gedanken, Journal of Applied Polymer Science, 104(3), 1423 (2007).
18. D.W. Chae, S.G. Oh, B.C. Kim, Journal of Polymer Science B: Polymer Physics, 42(5), 790 (2004).

Mater. Res. Soc. Symp. Proc. Vol. 1479 © 2012 Materials Research Society
DOI: 10.1557/opl.2012.1599

EVALUATION OF THE CAPACITY OF GRANULATION IN SURGICAL WOUNDS
WITH CONDENSED TANNINS IN MATRICES TiO2.

José Albino M. Rodriguez.[1], José Rutilio M. López[1], Genaro C. Gutiérrez[1], Marco-Antonio G Coronel[1], Enrique S. Mora[2], Lilián-Aurora M Rodríguez[3], Fernando M. Rodríguez[4]

[1] Facultad de Ciencias Químicas, [2]Instituto de Física, [3]Facultade Ciencias Físico-Matemáticas. Benemérita Universidad Autónoma de Puebla, Av. San Claudio y 14 Sur. Col. San Manuel, C. U. Puebla, Pue., C.P. 72570. [4] Hospital de Chiautla de Tapia, 11 Norte Carretera a Ixcamilpa S/N, C.P. 74730, Puebla.

ABSTRACT.

The nanoencapsulation in biocompatible inorganic materials with human cell activity is a leading technology to control the process of releasing the drug in the right place. At present, the sol-gel process has emerged as a promising platform for the immobilization, stabilization and encapsulation of biological molecules such as enzymes, antibodies, microorganisms, and a variety of drugs. The matrices obtained are chemically inert, hydrophilic and easy synthesis. They have high mechanical strength, thermal stability in wide temperature ranges and absorb organic solvents so insignificant compared with other organic polymers. They are resistant to microbial attack and exhibit high biocompatibility with the body, as provided for implantation in situ in the treatment of various diseases. An additional advantage is that it provides viability encapsulated molecules, since these matrices act as reservoirs of water thus helping to maintain the biological activity of enzymes, antibodies, cells, and drugs for the moisture level required for the molecule. We used the action of the active ingredients of tepezcohuite (condensed tannins) to assess the capacity aseptic surgical wound healing than 1 cm in diameter in New Zealand white rabbits. Experimentally and statistically demonstrating the effectiveness of healing nanoreservoirs Tan/TiO$_2$-150 the weight of tannins by 60% compared to condensed tannins as such, TiO$_2$ and isotonic saline.

Keywords: nanoscale, biomedical, sol-gel

Corresponding author: albinomx@yahoo.com

INTRODUCTION.

For 10 years, have developed alternative methods of application to improve the outcome of burn injuries as the application of gel hydroactive, occlusive hydrocolloid dressing with ionic silver hydrofiber, stretch bands, among others, however, most of them are inappropriate for presenting a very slow healing activity or by its high cost, resulting in a heavy burden on health services and patient [1, 2]. Mexico has numerous species of plants and trees used for therapeutic purposes including tepezcohuite tree, root name of the Nahuatl language that comes from the word tepezcuahuitl, meaning "tree from hill bleeding"[3, 4]. Anton et. al. prove that the fibers, starches, triterpenoids, saponins and condensed tannins, improve treatment for burns and skin regeneration. The latter, once extracted from the bark of tepezcohuite is a reddish brown powder, bitter and odorless. So that the tannins of Mimosa tenuiflora have a potent inducing factor and regeneration epidermal are especially effective for treating burns [5-8]. J. Fournairet. al.; the

Industrial Laboratory of Biology of Paris; the Laboratory Safepharm Ltd, London England and the Mexican Social Security Institute studies on the bacteriostatic activity, physiological and use in cosmetology to the tepezcohuite, they conclude that the crust of tepezcohuite, can be used widely in burns, demonstrating that has no side effects and having an antimicrobial effect and can also be used on infected wounds and open. The trauma and orthopedic José Velásquez Mijangos report on the great benefits of using tepezcohuite in patients with burns, wounds, fractures, and varicose ulcers [9-14]. The nanoencapsulation in arrays of inorganic materials biocompatible with the human body such as titanium oxide, it is systems "controlled" for to the stabilized, increase and eliminate side effects caused by systemic administration of drugs through the circulatory system. T. López et. al. studied, developed and implemented a number of nanoreservorios of TiO_2 doped with valproic acid (VPA) and phenytoin sodium phenyl (PSP) in the treatment of neurological disorders that occur primarily in the central nervous system. This ceramic device is implanted directly around the damaged tissue, to avoid passage through the blood brain barrier and reducing the dose of medication needed. Valproic acid (VPA), a commonly used anticonvulsant drug, encapsulated within a TiO_2 device has been successfully tested in mice as a therapy for seizure disorders [15-18]. In this project we synthesized nanoreservorios of TiO_2 by the encapsulation of active substances such as condensed tannins extracted from the mimosa tenuiflora (tepezcohuite). The nanoreservorios obtained are applied to surgical wounds of the type non-chronic shallow on the backs of New Zealand rabbits. Preliminary results show an evolution of wound closure in 90% with respect to Tan/TiO_2-150 nanoreservoirs, of the isotonic saline, to the tannins and to the nanomaterials of TiO_2.

EXPERIMENT.

We synthesize three nanoreservoirs in independent form through the sol-gel technique. TiO_2 (reference) and two that containing different concentrations of condensed tannins (50 mg and 150 mg in weight). They are labeled as Tan/TiO_2, Tan/TiO_2-50 and Tan/TiO_2-150 respectively. The synthesis was performed with the aid of a system reflux with constant stirring at 70°C of reaction temperature. All the nanoreservoirs were synthesized under the same conditions separately. The only variable was the mass concentration of condensed tannins in mg. All the nanoreservoirs were characterized with the help of UV-VIS spectrophotometer Varian Cary model 100, which has an integrating sphere coupled of diffuse reflectance and IR spectrophotometer with Fourier transform brand Scalibur Varian Digilab model. The efficiency of the TiO_2 and Tan/TiO_2-150 nanoreservoirs to demonstrate in the evolution in the granulation (healing) of the aseptic surgical wounds of 1 cm in diameter made to 6 New Zealand rabbits clinically healthy with an average weight of 3 kg.

Synthesis of the TiO_2 Nanoreservoirs.

To a previously prepared solution with 150 mL anhydrous isopropyl alcohol 99.5% from Sigma-Aldrich, 10 mL of deionized water and 1 g of polyvinylpyrrolidone of 55,000 amu (PVP-55000, Sigma-Aldrich) was added drop wise 19.2 mL of titanium isopropoxide (97%, Sigma-Aldrich) for a time of 3 hours in a reflux system at 70 ° C with constant agitation. The final solution property of gel is immersed in ice bath for 15 min to 3°C. The solvent is removed in a rotaryEseveD402-2 at 50°C in an integrated vacuum.TheTiO_2 nanoreservoir dried at 50°C in an oven RiossaH-33for 24 hours. The process of the nanoencapsulation condensed tannins in arrays of the titanium oxide, with a concentration of 50 and 150 mg of active substance (Tan/TiO$_2$-50, Tan/TiO$_2$-150) was similar to the synthesis of the TiO$_2$ nanoreservoir. The variant is the incorporation into corresponding mass (50 and 150 mg) of condensed tannins to the homogeneous solution in form separately.

DISCUSSION.

UV-VIS Spectroscopy

The optical and electronic properties of the TiO$_2$, Tan/TiO$_2$-150 and Tan/TiO$_2$-50 nanoreservoirs, we can they studied by the value of the band gap energy. The organic substances such as condensed tannins and tepezcohuite, we can only it identify the maximum absorption spectrum and the location of the wavelength in the UV-VIS region, however, we can relate these results with nanosemiconductors to study the change in the physicochemical properties of the active substance (condensed tannins) in TiO$_2$. The nanomaterial TiO$_2$ has a band gap magnitude of 371.2 nm, while the nanoreservorios Tan/TiO$_2$-150 and Tan/TiO$_2$-50 has a value of 401.4 nm and 427 nm respectively. It is proposed that a higher concentration of tannin in the matrix of TiO$_2$, the magnitude of band gap of the titanium oxide tends to move to the reference value of the nanomaterial. We conclude that the concentration by weight of condensed tannins clearly modifies the optical and electronic properties of TiO$_2$ as shown in Table I. It shows some optical and electronic properties of the nanoreservoirs.

Table I. Optical and Electronic Properties of TiO$_2$, Tan/TiO$_2$-50 and Tan/TiO$_2$-150 Nanoreservoirs.

Nanomaterial	λ (nm)	E_g (eV)	υ (Hz)(x10^{14})	Spectral region
Ti(nbut)$_4$-98%	349.0	3.55	8.60	Near UV
TiO$_2$	371.2	3.34	8.08	UV-VIS
Tan/TiO$_2$-50	427.0	2.90	7.02	VIS (violet)
Tan/TiO$_2$-150	401.4	3.09	7.47	UV-VIS (violet)

The optical properties of the nanoreservoir Tan/TiO$_2$-150 has a maximum absorption of radiation within the limits of ultraviolet-visible spectrum (color violet in the visible region of the electromagnetic spectrum), and for the nanoreservoir Tan/TiO$_2$-50 the maximum absorption occurs in the visible spectral region. The photon absorption produces certain electronic transitions originating from the highest state of the valence band to the lowest state of the conduction band. From the point of view of ligand field theory the electronic transition corresponds to the transition from non-electronic link (η) and the electronic state Pi (π), the final state of antibonding Pi (π^*). From the viewpoint of molecular orbital, the upper edge of the valence band is composed of oxygen 2p orbital, which form link p orbital, while the bottom of the conduction band, the bands are mainly formed by 4d atomic orbital of Ti.

FTIR Spectroscopy.

The FTIR spectrum of the bark of tepezcohuite shown the asymmetric stretching vibration modes (ν_{CH}) of the methyl (CH$_3$-) product of the component substances of the bark of tepezcohuite as the starch, steroids, flavanoids, resins, dextrin, glycerin and salicylic acid, are located in the region of 2942 cm^{-1} and type bending mode (v_{CH}) of C-H$_2$ species between 1321 cm^{-1} and 1329 cm^{-1}.In 1641 cm-1 mode presents of the symmetric vibration (v_{COO-}) of the carboxylate ion C=O, corresponding of to flavonoids and of the salicylic acid of the tepezcohuite. The functional groups of the aromatic vibration modes ($v_{Benzene-H}$) of the deformation outside the plane of the ring are located approximately in 567 cm^{-1}. In 3375 cm-1 is show the interactions of v_{OH} vibration modes of elongation corresponding to the species OH$^-$. In the FTIR spectrum condensed tannins are shown the stretching vibration modes (v_{C-C}) of the C-C groups, which correspond to the benzene ring annular skeleton in 1427 cm^{-1} and 1593 cm^{-1}. Vibration spectral bands of the aromatic compounds are located between the 554 cm^{-1} and 825 cm^{-1}. These represent the vibration modes of the deformation type ($v_{Benzene-H}$) of the phenyl-H species located outside the plane of the benzene ring. The Vibration modes bending located at 1023 cm-1 and 1161 cm-1 corresponding to (v_{R-O-R}) C-O-C groups. The v_{C-H} bending vibration modes of the species CH$_2$ located in 1279 cm^{-1} and the stretching vibration modes (v_{OH}) of the OH species are at 3285 cm^{-1}. The FTIR spectrum of the Tan/TiO$_2$-50 and Tan/TiO$_2$-150 shown the stretching and the bending mode vibration similar that the tepezcohuite and condensed tannins as v_{O-H}, v_{C-H}, δ_{HOH}, v_{COO-}, δ_{CH3}, and v_{C-C}. The vibration bands between 823 and 592 cm-1, are assigned to the vibration modes of bending type ($vTi-O$) of Ti-O groups. In this region of the infrared spectrum, there are the modes of vibration type vBenzene-H deformation outside the plane of the ring [x-xx]. It show figure 1.

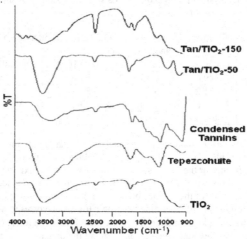

Figure 1. FTIR Spectra of the TiO$_2$, Tepezcohuite, tannins, Tan/TiO$_2$-50 and Tan/TiO$_2$-150

EVALUATION OF TiO$_2$ AND Tan/TiO$_2$-150 NANORESERVOIRS.

The efficiency of TiO$_2$ and Tan/TiO$_2$-150 nanoreservorios demonstrated in the evolution of granulation (closure) in aseptic surgical wounds of 1 cm in diameter made to 6 New Zealand rabbits clinically healthy between 2 to 3 kg, anesthetized intravenously into the marginal ear vein with sodium pentobarbital, 1 mL/2.5 kg. Determining the area of wound closure was performed using aseptic surgical aid of cellophane paper and marker. The edges of the wound were defined at intervals of two days; the cellophane is placed on graph paper to quantify how many squares occupied by each circle measured in mm^2. The rabbits were divided into treatments A, B and C. The process of wounds is the shaving of the dorsal part of each rabbit. To each rabbit was performed 6 intentional wounds previously using disinfected scissors in the dorsal area (back), 3 left and 3 on the right. Wounds located on the dorsal side of each rabbit was applied 2 or 3 drops (0.5 mL) of isotonic saline solution (ISS), which we will call them the wounds of control and the right side of the back (which we will call case) 30 mg was applied: condensed tannins (rabbits 1, 2: treatment A), nanoreservorio Tan/TiO$_2$-150 (rabbits 3, 4: treatment B) and TiO$_2$ (rabbits 5, 6: treatment C). Wounds from each treatment were checked daily for 15 days. At the thirteenth day we note that the nanoreservorio Tan/TiO$_2$ (treatment B) shows an evolution of granulation (close) respects to surgical wounds: condensed tannins (treatment A), the nanoreservorio TiO$_2$ (treatment C) and isotonic saline solution (I.S.S.), as shown in figure 2. Experimental data of measurements of the area of surgical wound closure for each treatment are shown in table II.

Table II. Comparative table of the closing speed of the six rabbits, measured in mm^2.

Day	Rabbit 1		Rabbit 2		Rabbit 3		Rabbit 4		Rabbit 5		Rabbit 6	
	ISS	Tan	ISS	Tan	ISS	Tan/TiO$_2$	ISS	Tan/TiO$_2$	ISS	TiO$_2$	ISS	TiO$_2$
					Closingarea mm^2							
0	25.0	25.2	26.7	26.3	28.7	26.7	26.3	25.7	24.3	25.0	21.0	23.7
1	24.5	24.7	25.3	24.1	27.1	24.1	24.6	23.2	22.7	23.7	19.6	22.1
2	21.8	22	22.3	19.7	24.6	19.7	22.4	19.8	19.6	19.9	16.5	19.8
3	15.7	16	17.8	15.3	21	15.3	16	13.7	16.7	17.3	13.3	15
4	14.5	14.6	16.1	13.7	17.2	13.7	15.2	12.1	15.1	15.0	12.5	13.7
5	12.3	12.8	14.6	11.4	13.8	11.4	14.6	9.9	14.7	13.4	10.4	11.1
6	11	10.8	12.4	8.7	11.7	8.7	13.7	8.3	13.3	12.4	8.3	10.7
7	9.7	8.3	8.9	6.8	9.5	6.8	10.3	5.7	12.6	11.7	6.7	8.8
8	8.0	7.4	6.1	4.1	8.0	4.1	8.0	3.7	11	10.8	5.7	7.3
9	6.7	6.1	4.8	2.7	5.8	2.7	5.9	1.5	9.5	8.7	4.9	5.8
10	5.6	5.0	3.5	0.6	3.5	0.6	3.6	0.8	7.3	6.4	3.1	3.5
11	4.9	4.5	2.7	0.0	2.4	0.2	2.0	0.0	5.1	4.0	2.8	2.5
13	1.9	1.5	1.3	0.0	1.3	0.0	1.5	0.0	3.0	2.0	1.5	1.7
14	0.4	0.8	0.4	0.0	0.4	0.0	0.6	0.0	2.5	1.2	0.3	0.9
15	0.0	0,0	0.0	0.0	0.0	0.0	0.0	0.0	1.1	0.0	0.0	0.3

CONCLUSION.

The nanoreservorios of Tan/TiO$_2$-150 show a faster closure of surgical wounds caused intentionally on the back of the rabbits, the evolution of healing (closure) of these were presented at 11 days after having caused the injuries. Tannins extracted from the bark of tepezcohuite show a speed of healing at 13 days, but healing is not complete. The reference nanoreservorio TiO$_2$ showed similar activity to the tannins. While isotonic saline showed a partial healing at 13 days, where cicaticación is partial. Therefore, according to results nanoreservorios efficiency in the speed of healing (closure) provides the following scheme: Tan/TiO$_2$-150> $\geq$ condenzados tannins TiO2> isotonic saline. Allotropic phase of TiO$_2$ in which are encapsulated condensed tannins tends to rutile, whose value is 3.09 eV Eg, while anatase TiO$_2$ tends to be distorted (3.3 eV).

REFERENCES.
1. D. M. Zapata, A. Estrada, "Calidad de vida relacionada con la salud de las personas afectadas por quemaduras después de la cicatrización", Medellín, Colombia, Biomédica **30**, 492-500 (2010).
2. A. Cuadrac, J.L. Pineros, R. Roa, Rev. Med. Clin. CONDES **21** (1), 41-45 (2010).
3. S.L. Camargo-Ricalde, "Descripción, distribución, anatomía, composición química y usos de Mimosa tenuiflora (Fabaceae-Mimosoideae) en México", Rev. biol. Trop, **48** 4 (2000).
4. B. Morales, J.A. Pérez, "Manual de identificación de árboles de selva baja mediante cortezas", Cuadernos No. 6. Inst. Biología, UNAM, 83 México (1990).
5. J.R. Navarro. "Propiedades farmacológicas y extracción de principios activos de la corteza del tepezcohuite". Tesis de licenciatura, UAM-I, marzo (1988).
6. R. Anton, Y. Jiang, B. Weniger, J. P. Beck and L. Rivier, "Pharmacognosy of Mimosa tenuiflora, (Willd.) Poiret." J. Ethnopharmacol. **38**, 153-157 (1993).
7. M. M. Lozoya, X. Lozoya and J.L. González, "Propiedades farmacológicas *in vitro* de algunos extractos de Mimosa tenuiflora (tepescohuite)". Arch. Invest. Méd, **21**, 163-169 (1990).
8. A. Urióstegui, "Algunas plantas curativas de Taxco, Guerrero, México", Tlahui-Medic, **2**, 26 (2008).
9. J. Founai, "Actividad Bacteriológica, Fisiológica y su uso en Cosmetología del Tepezcohuite". Facultad de Farmacia. Universidad del sur de París, **1** (1990).
10. J. Founai, Facultad de Farmacia. "Reporte de la evaluación de la actividad bacteriostática del tepezcohuite". Universidad de París **1** (1991).
11. Laboratorio Industrial de Biología, "Análisis General del Tepezcohuite", París, Francia. (1991).
12. Safepharm Laboratories Ltda, "Examen de Toxicidad del Tepezcohuite", Londres, Inglaterra, **1** (1991).
13. Instituto Mexicano del Seguro Social, "Reporte sobre el Efecto Antimicrobiano de los Extractos de la Corteza del Tepezcohuite", **1**, (1991).
14. M. R. Macédo., S. Vieira Pereira, L. Filgueiras, A. Vieira, O. Guedes, "Atividade Antimicrobiana e Antiaderente do Extrato da Mimosa tenuifl ora (Willd). Poir. Sobre Microrganismos do Biofilme Dentário", Pesq. Bras. Odontoped Clin. Integr. João Pessoa, **9** 2, 161-165 (2009).
15. T. López, E. Ortiz-Islas, E. Vinogradova, J. Manjarrez, J.A. Azamar, J. J. Alvarado-Gil and P. Quintana, "Optical Materials Structural, Optical andVibracional Properties of Sol-Gel Titania Valproic Acid Reservoirs", **29**, 82–87 (2006).
16. T. López, E. I. Basaldella, M.L. Ojeda, J. Manjarrez and R. Alexander-Katz, "Optical Materials Encapsulation of valproic acid and sodicphenytoine in ordered mesoporous SiO₂ solids for the treatment of temporal lobe epilepsy", **29**, 75–81 (2006).
17. T. López, J. Manjarrez, D. Rembao, E. Vinogradova, A. Moreno, R. D. González, "An implantable sol-gel derived titania-silica carrier system for the controlled release of anticonvulsants", Mat. Lett **60**, 2903-2908, (2006).
18. A. Peterson, T. López, E. Ortiz-Islas and R.D. González, "Pore structures in an implantable sol-gel Titania ceramic device used in controlled drug release applications: A modeling study", Applied Surface Science **253**, 5767–5771, (2007).

Mater. Res. Soc. Symp. Proc. Vol. 1479 © 2012 Materials Research Society
DOI: 10.1557/opl.2012.1600

Cyclotriveratrylene Dendrimers

Sanchez-Montes Karla E.,[1] Klimova Tatiana,[2] Martínez-Klimov Mark E.,[2]
Martínez-García Marcos.[1*]

[1]Instituto de Química, Universidad Nacional Autónoma de México, Cd. Universitaria,
Circuito   Exterior,   Coyoacán,   C.P.   04510,   México   D.F.   E-mail:
margar@servidor.unam.mx
[2]Facultad de Química, Universidad Nacional Autónoma de México, Cd. Universitaria,
Circuito Interior, Coyoacán, C.P. 04510, México D.F.

ABSTRACT

Dendrons with dodecyl ended groups joined benzyloxy moieties were attached to a
cyclotriveratrylene core. The dendrimers were used in Diederich cyclopropanation
reaction with the fullerene C_{60}. The structure of the synthesized dendrimers was
confirmed by ^{1}H- and ^{13}C-NMR, MALDI-TOF mass spectrometry and elemental
analysis.

Keywords: Cyclotriveratrylene; dendrimers; fullerene C_{60}

INTRODUCTION

The cup-shape cyclotriveratrylene derivatives (CTV) are among the key structures,
they have been exploited for many years for the design of molecular hosts[1]. They still
represent interesting starting compounds for the construction of new macromolecular
architectures such as solid inclusion complexes[2], biosensors, chiroptical properties[3]
chiral scaffolds for triple helix formation[4], organo- or metallo-gels[5], H_2 storage
targets[6] or more recently, coordination polymer networks[7] and self- assembled
monolayers on gold surface[8] and dendrimers[9]. The tree or six p-hydroxybenzyl
substituent present on the wider rim of cyclotriveratrylenes are able suffer
modification to obtain dendrimers without steric restrictions. Now we report the
synthesis of cyclotriveratrylene-dendrimers with three and six fullerenes in the
dendritic branches.

EXPERIMENT

Materials and equipments

Solvents and reagents were purchased as reagent grade and used without further
purification. Acetone was distilled over calcium chloride. Tetrahydrofuran was
distilled from sodium and benzophenone. Column chromatography was performed on
Merck silica gel 60Å (70-230 mesh). ^{1}H- and ^{13}C-NMR were recorded on a Varian-
Unity-300 MHz with tetramethylsilane (TMS) as an internal reference. Infrared (IR)
spectra were measured on a spectrophotometer Nicolet FT-SSX. Elemental analysis
was determined by Galbraith Laboratories Inc. (Knoxville, TN, USA). FAB+ mass
spectra were taken on a JEOL JMS AX505 HA instrument. Electrospray mass spectra
were taken on a Bruker Daltonic, Esquire 6000. MALDI-TOF mass spectra were
taken on a Bruker Omni FLEX.
Synthesis of dendrimers

A mixture of **12** or **14** (0.36 mmol), cesium carbonate (0.36 mmol) in N,N-dimethylformamide anhydrous (8 ml) was heated to reflux and stirred vigorously under nitrogen after 20 min. The cyclotriveratrylene **3** (0.07 mmol) dissolved in N,N-dimethylformamide anhydrous (10 ml) was added dropwise and the reaction was continued for 31 h at reflux. The mixture was allowed to cool and the precipitate was filtered. The filtrate was evaporated to dryness under reduced pressure and the residue was chromatographed on silica gel with ethyl hexane as the eluent to afford the cyclotriveratrylene-dendrimer.

Dendrimer 15. Yield (0.25 g, 0.13 mmol, 42 %), dark-red glassy. UV (CHCl$_3$, nm): 284, 247. IR (Film, cm^{-1}): 2925, 2854, 1741, 1600, 1461, 1253, 1166, 1064, 834, 721. ^{1}H NMR (CDCl$_3$, 400 MHz): δ= 0.88 (t, 36H, J= 6.1 Hz), 1.26 (m, 216H), 1.70 (m, 24H), 3.45 (s, 12H), 3.50 (d, 3H, J= 13.0 Hz), 3.80 (t, 24H, J= 5.3 Hz), 3.84 (s, 9H), 4.61 (s, 6H,), 4.60 (d, 3H, J= 12.0 Hz), 5.04 (s, 12H,), 5.08 (s, 12H,), 6.35-6.60 (m, 27H,), 7.54 (s, 3H,), 7.70 (s, 3H,). ^{13}C NMR (CDCl$_3$): δ= 14.1 22.7, 26.1, 29.4, 29.7, 32.0, 36.3, 44.8, 56.1, 68.0, 70.2, 71.6, 100.6, 101.1, 104.9, 106.2, 113.6, 115.5 (Ar-, 131.2, 132.4, 137.2, 141.3, 143.8, 144.2, 145.2, 160.4, 162.0; MALDI-TOF-MS: m/z 4124 [M]$^+$. Anal. Calcd. for C$_{255}$H$_{390}$O$_{42}$: C, 74.20, H, 9.52 %. Found: C 74.18, H 9.52.

Dendrimer 16. Yield (0.33 g, 0.10 mmol, 31 %), dark-brown glassy. UV (CHCl$_3$, nm): 283, 243. IR (Film, cm^{-1}): 3422, 2925, 2854, 1740, 1599, 1461, 1244, 1166, 1063, 833. ^{1}H NMR (CDCl$_3$, 300 MHz): δ= 0.89 (t, 72H, J= 6.3 Hz), 1.27 (m, 432H), 1.74 (m, 48H), 3.48 (s, 24H), 3.51 (d, 3H, J= 12.6 Hz), 3.91 (s, 9H), 3.94 (t, 48H, J= 6.3 Hz), 4.62 (s, 6H), 4.84 (d, 3H, J= 11.2 Hz), 5.02 (s, 36H), 5.05 (s, 24H), 5.10 (s, 6H), 6.41 (t, 12H, J= 2.1 Hz), 6.46 (d, 18H, J= 2.4 Hz), 6.48 (d, 24H, J= 2.1 Hz), 6.51 (t, 9H, J= 3.0 Hz), 7.60 (s, 3H), 7.61 (s, 3H). ^{13}C NMR (CDCl$_3$): δ= 14.1, 23.0, 26.1, 29.4, 29.6, 32.0, 36.5, 46.4, 56.1, 66.3, 68.1, 100.6, 101.0, 101.4, 105.1, 106.5, 107.0, 112.3, 115.6, 132.7, 133.8, 137.2 , 137.9 , 139.3 , 143.2 , 145.2 , 160.5 , 170.8: MALDI-TOF-MS: m/z 8213 [M]$^+$. Anal. Calcd. for C$_{507}$H$_{774}$O$_{84}$: C, 74.14, H, 9.50 %. Found: C 74.16, H 9.51 %.

Synthesis of fullerene derivatives

DBU (0.43 mmol) and I$_2$ (0.020 mmol) were added at room temperature to a solution of C$_{60}$ (0.17 mmol) and **15** or **16** (0.17 mmol) in 17 ml of toluene anhydrous. The mixture was stirred for 3 days. The mixture was filtered and evaporated under vacuum and purified by chromatographic column (SiO$_2$, CH$_2$Cl$_2$/methanol, 8:2).

Fullerene derivative 17: Yield (0.06 g, 0.01 mmol, 26 %), dark-brown solid. UV (CHCl$_3$, nm): 248. IR (Film, cm^{-1}): 3258, 2925, 2854, 1733, 1648, 1600, 1459, 1292, 1164, 1067, 835, 751. ^{1}H NMR (CDCl$_3$, 300 MHz): δ= 0.85 (t, 36H, , J= 6.1 Hz), 1.23 (m, 216H,), 1.70 (m, 24H,), 3.50 (d, 3H, , J= 13.0 Hz), 3.80 (t, 24H, , J= 5.3 Hz), 3.84 (s, 9H,), 4.61 (s, 6H,), 4.60 (d, 3H, , J= 12.0 Hz), 5.04 (s, 12H, -), 5.08 (s, 12H, -), 6.35-6.60 (m, 27H,), 7.54 (s, 3H,), 7.70 (s, 3H,). ^{13}C NMR (CDCl$_3$): δ= 14.2 , 22.8 (), 26.1 (), 26.2 (), 29.4 (), 29.8 (), 32.0 (), 37.3 (), 56.3, 66.8 (C$_{60}$), 67.0 (C$_{60}$), 68.1 (), 69.5 (C$_{60}$), 100.6 , 101.0 , 101.1 , 101.5 , 105.1 , 106.3 , 106.5 , 125.1 (C$_{60}$), 126.7 (C$_{60}$), 126.7 (C$_{60}$), 128.9 (C$_{60}$), 130.9 (-), 132.4 (-), 133.8 (C$_{60}$), 135.3 (C$_{60}$), 136.7 , 137.5 (C$_{60}$), 137.6 (C$_{60}$), 138.1 (C$_{60}$), 138.2 (C$_{60}$), 138.4 (C$_{60}$), 139.0

(C_{60}), 139.3 (C_{60}), 140.7 (C_{60}), 141.3 (C_{60}), 142.2 (C_{60}), 142.6 (C_{60}), 143.4 , 143.5 (C_{60}), 143.9 (C_{60}), 144.2 (, C_{60}), 144.6 (C_{60}), 144.7 (C_{60}), 144.8 (C_{60}), 145.4 (, C_{60}), 145.6 (C_{60}), 145.7 (C_{60}), 145.8 (C_{60}), 145.9 (C_{60}), 146.3 (C_{60}), 146.4 (C_{60}), 146.6 (C_{60}), 147.0 (C_{60}), 147.3 (C_{60}), 147.8 (C_{60}), 148.1 (C_{60}), 155.2 , 160.4 , 165.5. MALDI-TOF-MS: m/z 6295 [M+Na]$^+$. Anal. Calcd. for $C_{435}H_{378}O_{42}$: C, 83.23, H, 6.07 %. Found: C 83.25, H 6.09 %

Fullerene derivative 18: Yield (0.06 g, 0.01 mmol, 26 %), red-brown solid. UV (CHCl$_3$, nm): 247. IR (Film, cm^{-1}): 3422, 2925, 2854, 1740, 1599, 1461, 1244, 1166, 1063, 833, 721. ^{1}H NMR (CDCl$_3$, 300 MHz): δ= 0.89 (t, 72H, , J= 6.3 Hz), 1.27 (m, 432H,), 1.74 (m, 48H,), 3.49 (d, 3H, , J= 14.0 Hz), 3.85 (s, 9H,), 3.94 (t, 48H, , J= 6.3 Hz), 4.62 (s, 6H,), 4.70 (d, 3H, , J= 13.2 Hz), 5.02 (s, 36H,), 5.05 (s, 24H,), 5.10 (s, 6H,), 6.41 (t, 12H, , J= 2.1 Hz), 6.46 (d, 18H, , J= 2.4 Hz), 6.48 (d, 24H, , J= 2.1 Hz), 6.51 (t, 9H, , J= 3.0 Hz), 7.60 (s, 3H,), 7.61 (s, 3H,). ^{13}C NMR (CDCl$_3$): δ= 14.0 , 22.7 (), 26.1 (), 26.2 (), 29.4 (), 29.7 (), 32.0 (), 38.0 (), 56.1, 66.1 (C_{60}), 66.5 (C_{60}), 68.3 (), 69.5 (C_{60}), 100.6 , 100.7 , 101.0 , 101.1 , 101.5 , 105.5 , 105.8 , 106.1 , 106.3 , 113.2 , 114.0 (H), 125.3 (C_{60}), 126.5 (C_{60}), 127.0 (C_{60}), 128.8 (C_{60}), 131.4 (-), 132.3 (CTV-), 133.4 (C_{60}), 135.2 (C_{60}), 136.6 , 137.0(C_{60}) , 137.3 (C_{60}), 138.3 (C_{60}), 138.5 (C_{60}), 139.6 (C_{60}), 139.7 (C_{60}), 140.4 (C_{60}), 141.7 (C_{60}), 142.2 (C_{60}), 143.1 (C_{60}), 143.6 (C_{60}), 143.7 , 143.8 (C_{60}), 143.9 (C_{60}), 144.2 (, C_{60}), 145.3 (C_{60}), 145.4 (C_{60}), 145.5 (, C_{60}), 145.6 (C_{60}), 145.7 (C_{60}), 145.8 (C_{60}), 145.9 (C_{60}), 146.0 (C_{60}), 146.5 (C_{60}), 147.2 (C_{60}), 147.3 (C_{60}), 147.8 (C_{60}), 148.1 (C_{60}), 155.2 , 158.2 , 158.4. MALDI-TOF-MS: m/z 12513 [M]$^+$. Anal. Calcd. for $C_{867}H_{750}O_{84}$: C, 82.20, H, 6.04 %. Found: C 83.30, H 6.06 %.

DISCUSSION

The cyclotrivetrylene derivative **3** in ca. 47 % was obtained in agreement with the literature data[10].

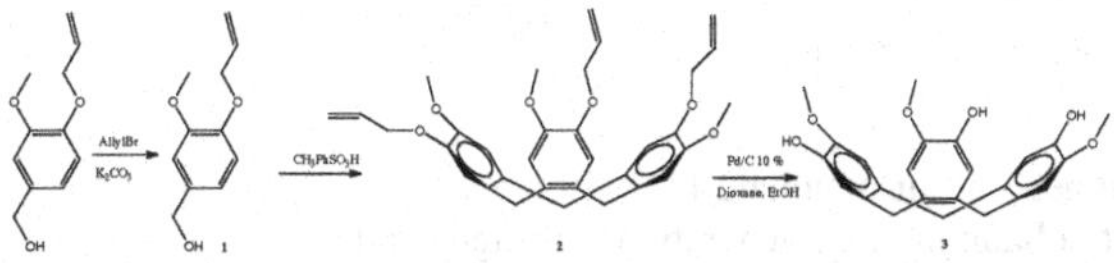

Scheme 1. Synthesis of cyclotriveratrylene (a) K$_2$CO$_3$, acetone, reflux; (b) (CH$_3$)PhSO$_3$H; (c) Dioxane Pd/charcoal, ethanol, perchloric acid.

Dendrons containing polybenzyl ether groups were prepared in agreement with the literature data[11] (Scheme 2). The compound **9** was obtained from the diethyl 5-(hydroxymethyl)isophthalate **7**, which first was protected with *tert*-butylchlorodimethylsilane to obtain the compound **8**, followed by the reduction with LiAlH$_4$ in THF to give dimethanol **9** (Scheme 2).

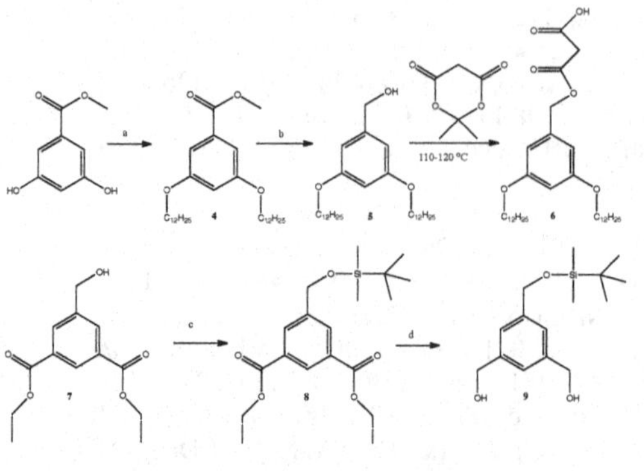

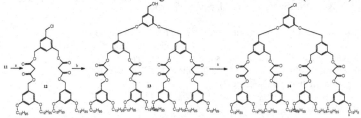

Scheme 2. Synthesis of **4** and **7** a) $C_{12}H_{25}Br$, K_2CO_3, KI, acetone; c) THF, LiAlH$_4$, 0°C; c) TBSCl, imidazole, THF, 0°C; d) THF, LiAlH$_4$, 0°C.

The compound **6** was attached to the compound **9** to obtain the bismalonate **10** in CH$_2$Cl$_2$ with DCC and DMAP. The bismalonate **10** was treated with 1.0 M tetrabutylammonium fluoride to obtain the bismalonate **11** (Scheme 3).

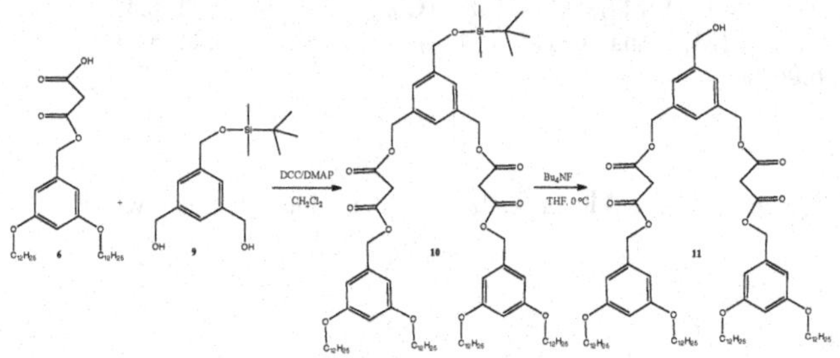

Scheme 3. Synthesis of bismalonate **11**

The bismalonate **11** was used to obtain the first generation activated dendron **12** upon treatment with SOCl$_2$, which when coupled to the 3,5-dihydroxybenzyl alcohol was used to obtain the second generation activated dendron **14** (Scheme 4).

Scheme 4. Synthesis of activated dendrons **12** and **14**; a) SOCl$_2$, pyridine, CH$_2$Cl$_2$ b) 3,5-dihydroxybenzyl alcohol, Cs$_2$CO$_3$, DMF

The dendrons of first and second generation **12** and **14**, respectively, were attached to the cyclotriveratrylene **3** to obtain the dendrimers **15** and **16** (Scheme 5). The reaction was carried out in DMF and Cs$_2$CO$_3$ at reflux for 2 days, and the dendrimers **15** and **16** were obtained in good yields.

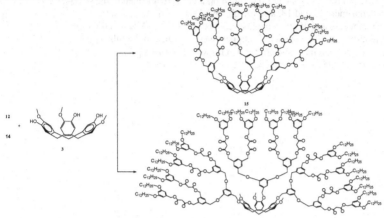

Scheme 5. Synthesis of dendrimers **15** and **16**.

The functionalization of C$_{60}$ was made using the double *Bingel* addition [11]. Treatment of C$_{60}$ with dendrimers **15** or **16**, I$_2$, and 1,8-diazabicyclo[5.4.0]undec-7-ene (DBU) in toluene at room temperature gave the fullerene derivatives **17** and **18** in 21 and 18 % yield, respectively (Scheme 6). All the spectroscopic studies and elemental analysis results were consistent with the proposed molecular structures assigned to the fullerene derivatives.

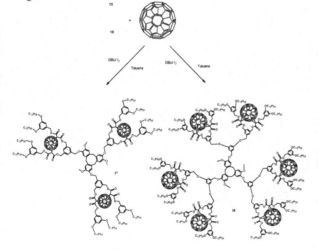

Scheme 6. Synthesis of fullerene derivatives **17** and **18**.

73

In the ^{1}H NMR spectra of compounds **17** and **18** (figure 1, 1a and 1b respectively) are observed; at δ_H 0.85-1.70 and 3.80 signals due to the aliphatic chain. Two doublets at δ_H 3.50 and 4.60 assigned to the $-CH_2-$ groups of the cyclotriveratrylene. Signals at δ_H 4.61, 5.04, and at δ_H 5.08 were ascribed to the $-CH_2-$O protons. For dendrimers **17** and **18**, also are observed signals of the aromatic protons at the macrocycle and of the benzyloxy chains. In the ^{13}C NMR spectra, of the cyclotriveratrylene-fullerene-dendrimers 35 signals were ascribed to the fullerene moiety. In the MALDI-TOF mass spectra for dendrimer **17** and **18** a peak at m/z=6295 (M+Na) and 12513 respectively were observed (Figure 2).

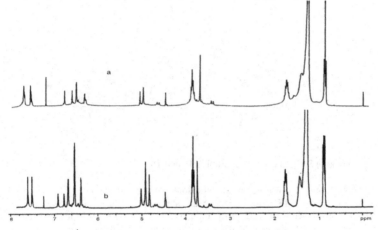

Figure 1. ^{1}H NMR spectra of dendrimers a) **17** and b) **18** in CDCl$_3$

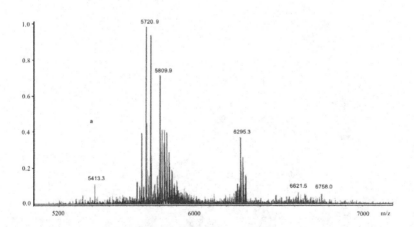

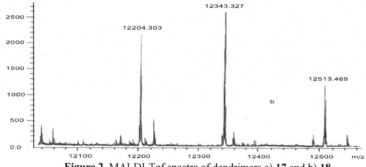

Figure 2. MALDI-Tof spectra of dendrimers a) **17** and b) **18**

3. CONCLUSIONS

The cyclotriveratrylene dendrimers were obtained from the dendrons **12** and **14**, and then the fullerene C_{60} was introduced in the dendritic branches by cyclization reaction. Adding the fullerene C_{60} in the last step of the synthesis gives good yields. The ^1H, ^{13}C NMR and UV-vis studies showed that the double Bingel cyclopropanation reaction afforded only *cis*-2 adducts in agreement with the literature data.

ACKNOWLEDGEMENTS

This work was supported by DGAPA-UNAM (IN-202010-3) grant. We would also like to thank Rios O. H., Velasco L., Huerta S. E., Patiño M. M. R., Peña Gonzalez M. A., and Garcia Rios E. for technical assistance.

REFERENCES

1. A. Collet, *Tetrahedron*, **43**, 5725 (1987).
2. M. J. Hardie, C. L. Raston, B. Wells, *Chem. Eur. J.*, **6**, 3293 (2000).
3. T. Brotin, J. P. Dutasta, *Chem. Rev.*, **109**, 85 (2009).
4. E. T. Rump, D. T. S. Rijkers, H. W. Hilbers, P. G. De Groot, R. M. J. Liskamp, *Chem. Eur. J.*, **8**, 4613 (2002).
5. D. Bardelang, F. Camerel, R. Ziessel, M. Schmutz, M. J. Hannon, *J. Mater. Chem.*, **18**, 489 (2008).
6. N. B. McKeown, B. Gahnem, K.J. Msayib, P.M. Budd, C. E. Tattershall, K. Mahmood, S. Tan, D. Book, H. W. Langmi, Walton, A. *Angew. Chem., Int. Ed.*, **45**, 1804 (2006).
7. M. J. Hardie, C. J. Sumby, *Inorg. Chem.*, **43**, 6872 (2004).
8. S. Zhang, L. Echegoyen, *J. Am. Chem. Soc.*, **127**, 2006 (2005).
9. J. F. Eckert, D. Byrne, J. F. Nicoud, L; Oswald, J. F. Nierengarten, M. Numata, A. Ikeda, S. Shinkai, A. Armaroli, *New J Chem.*, **24**, 749 (2000).
10. I. V. Lijanova, J. F. Maturano, J. G. D. Chavez, K. E. S. Montes, S. H. Ortega, T. Klimova, M. Martinez-Garcia, *Supramol. Chem.*, **21**, 21 (2009).
11. C. Bingel, *Chem. Ber.*, **126**, 1957 (1993).

Mater. Res. Soc. Symp. Proc. Vol. 1479 © 2012 Materials Research Society
DOI: 10.1557/opl.2012.1601

Nanostructured SBA-15 Materials as Appropriate Supports for Active Hydrodesulfurization Catalysts Prepared from HSiW Heteropolyacid

Mendoza-Nieto J.A., Tejeda-Espinosa K.D., Puente-Lee I., Salcedo-Luna C. and Klimova T.

Facultad de Química, Departamento de Ingeniería Química, Universidad Nacional Autónoma de México (UNAM), Cd. Universitaria, Coyoacán, México D.F., 04510, México.

E-mail: klimova@unam.mx

ABSTRACT

A series of NiW catalysts supported on SBA-15-type materials modified with Al, Ti or Zr were prepared and tested in simultaneous hydrodesulfurization (HDS) of two model compounds: dibenzothiophene (DBT) and 4,6-dimethyldibenzothiophene (4,6-DMDBT). Catalysts were prepared by incipient wetness impregnation of SBA-type materials (pure silica SBA-15, Al-SBA-15, Ti-SBA-15 or Zr-SBA-15) using Keggin-type heteropolyacid $H_4SiW_{12}O_{40}$ as active phase precursor and nickel nitrate. Nominal composition of the catalysts was 19 wt.% of WO_3 and 3 wt.% of NiO. The supports and catalysts were characterized by SEM-EDX, N_2 physisorption, small-angle and powder XRD, UV-Vis DRS, TPR and HRTEM. It was shown that a good dispersion of Al, Ti and Zr species on the SBA-15 surface was reached. The characteristic structure of the SBA-15 support was preserved in all supports and NiW catalysts. Addition of metal atoms (Al, Ti, Zr) on the SBA-15 surface prior to catalysts' preparation improved dispersion of Ni and W oxide species in calcined catalysts. HRTEM characterization of sulfided catalysts showed that the dispersion of NiW active phase was also better on metal-containing SBA-15 supports than on the pure silica one. All NiW catalysts supported on metal-containing SBA-15 materials showed an outstanding catalytic performance in HDS of both model compounds used (DBT and 4,6-DMDBT). A good correlation was found between the dispersion of sulfided NiW active phase and catalytic activity results. The highest HDS activity was obtained with the NiW catalyst supported on Zr-containing SBA-15 molecular sieve, which makes it a promising catalytic system for ultra-deep hydrodesulfurization of diesel fuel.

KEYWORDS: W, Catalytic materials, Nanostructure

INTRODUCTION

Nowadays, the current demand for high quality low sulfur transportation fuels is growing due to the necessity to solve environmental problems induced by SO_x emissions. Many efforts are aimed at designing more active or more selective hydrodesulfurization (HDS) catalysts, by either using new catalytic supports or changing the active phase [1]. Previous studies clearly show that the support nature and characteristics play an important role in the catalytic behavior of HDS catalysts. Well-ordered SBA-15 mesoporous molecular sieves, especially modified by the incorporation of different heteroatoms (Al, Ti, Zr, P, W, etc.), were found to be promising materials as supports for HDS catalysts [2-5]. Recently, heteropolyacids have attracted attention as precursors for HDS catalysts, because of their high solubility and particular molecular structure, in which different elements are associated in the same entity and can be deposited together from the same impregnation solution [6-8].

In the present work, we report results obtained from a series of NiW catalysts supported on SBA-15-type materials modified with Al_2O_3, TiO_2 or ZrO_2. Keggin-type tungstosilicic acid was used as a precursor for the catalyst preparation. Prepared samples were characterized and tested in simultaneous hydrodesulfurization (HDS) of two model compounds: dibenzothiophene (DBT) and 4,6-dimethyldibenzothiophene (4,6-DMDBT). The aim of this study was to inquire on the effect of the characteristics of the M-SBA-15 support (where M = Al, Ti, Zr) on the activity of NiW catalysts in deep HDS.

EXPERIMENTAL DETAILS

SBA-15 silica was synthesized according to well-known procedure [9]. Al-, Ti- or Zr-containing SBA-15 materials (M-SBA-15) were prepared by chemical grafting of corresponding metal salts or alcoxides ($AlCl_3$, $Ti(i-PrO)_4$ or $Zr(n-PrO)_4$) on the SBA-15 surface following experimental procedure described previously [4,5]. Pure silica SBA-15 and M-SBA-15 materials were used as supports for NiW catalysts. The catalysts were prepared by successive incipient wetness impregnation of methanol solutions of Keggin-type heteropolyacid $H_4SiW_{12}O_{40}$ (HSiW) as active phase precursor and nickel nitrate, $Ni(NO_3)_2 \cdot 6H_2O$. After each impregnation, the catalysts were dried (100 °C, 6 h) and calcined (500 °C, 4 h). The nominal metal loadings in the catalysts were 19 wt. % of WO_3 and 3 wt. % of NiO.

Supports and catalysts were characterized by SEM-EDX, N_2 physisorption, small-angle and powder X-ray diffraction (XRD), UV-vis diffuse reflectance spectroscopy (DRS), temperature-programmed reduction (TPR) and HRTEM. The HDS activity tests were performed in a batch reactor at 300 °C and 7.3 MPa total pressure for 8 h using hexadecane solution of DBT (1300 ppm of S) and 4,6-DMDBT (500 ppm of S). Before the activity tests, the catalysts were sulfided *ex situ* in a tubular reactor at 400 °C for 4 h. The course of the reaction was followed by withdrawing aliquots each hour and analyzing them by GC.

RESULTS AND DISCUSSION

Characterization of supports and catalysts

The chemical composition of the synthesized M-SBA-15 supports was determined by SEM-EDX. It was found that metal loadings were 8 wt. % of Al_2O_3, 13 wt. % of TiO_2 and 19 wt. % of ZrO_2 for Al-SBA-15, Ti-SBA-15 and Zr-SBA-15 supports, respectively. In the three cases, amount of metal (Al, Ti, Zr) loaded by chemical grafting on the SBA-15 surface was close to 2 metal atoms/nm^2, which is in line with previous publications [4,5]. Textural characterization of SBA-15 supports and corresponding catalysts (table I) showed that BET surface area, micropore area and total pore volume of SBA-15 material slightly decreased after the incorporation of Al, Ti, Zr and impregnation of Ni and W species on the supports. However, no significant differences were observed in the mesopore diameter (D_P) and pore wall thickness (δ) of the (M)-SBA-15 supports and NiW catalysts. This result points out that metal atoms (Al, Ti and Zr) incorporated to SBA-15, as well as Ni and W species, were well-dispersed.

Table I. Textural[a] and structural[b] characteristics of supports and NiW catalysts.

Sample	S_{BET} (m^2/g)	S_μ (m^2/g)	V_P (cm^3/g)	D_P (Å)	a_o (Å)	δ (Å)
SBA-15	707	56	0.98	78	110	32
Al-SBA-15	609	48	0.86	78	110	32
Ti-SBA-15	526	47	0.76	75	111	36
Zr-SBA-15	498	54	0.70	76	111	35
NiW/SBA-15	563	56	0.78	77	110	33
NiW/Al-SBA-15	490	62	0.67	77	110	33
NiW/Ti-SBA-15	429	61	0.60	75	110	35
NiW/Zr-SBA-15	422	62	0.59	75	110	35

[a] S_{BET}, specific surface area calculated by the BET method; S_μ, micropore area determined by the t-plot method; V_P, total pore volume; D_P, pore diameter corresponding to the maximum of the pore size distribution obtained from the adsorption isotherm by the BJH method.
[b] a_o, unit cell parameter determined from d_{100} distance as $a_o = d_{100} \times 2/\sqrt{3}$; δ, pore wall thickness calculated as $\delta = a_o - D_P$.

The nitrogen adsorption-desorption isotherms for the supports and catalysts are shown in figure 1. All isotherms correspond to type IV and exhibit an H1 hysteresis loop characteristic of SBA-15 nanostructured material. The sharpness of the adsorption branches is indicative of a narrow mesopore size distribution in all cases.

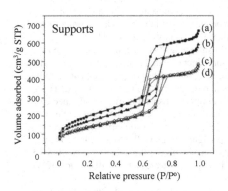 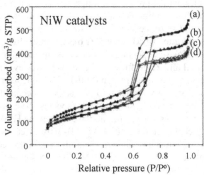

Figure 1. N_2 adsorption-desorption isotherms of supports: (a) SBA-15, (b) Al-SBA-15, (c) Ti-SBA-15, (d) Zr-SBA-15; and corresponding NiW catalysts.

Small-angle XRD patterns of the (M)-SBA-15 supports (figure 2, A) and corresponding NiW catalysts showed three reflections characteristic for *p6mm* hexagonal pore arrangement of the SBA-15 material. This long-ordered pore structure was preserved in all supports with grafted Al, Ti or Zr species and NiW catalysts. Powder XRD characterization confirmed a good dispersion of Ni and W oxide species in all prepared catalysts (figure 2, B).

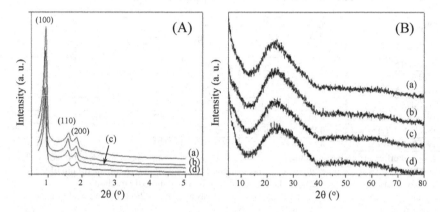

Figure 2. (A) Small-angle XRD patterns of supports: (a) SBA-15; (b) Al-SBA-15; (c) Ti-SBA-15 and (d) Zr-SBA-15. (B) Powder XRD patterns of catalysts: (a) NiW/SBA-15; (b) NiW/Al-SBA-15; (c) NiW/Ti-SBA-15 and (d) NiW/Zr-SBA-15.

Reduction profiles of NiW/(M)-SBA-15 catalysts are presented in figure 3 and corresponding degrees of reduction (α) in table II. Exact assignment of the TPR signals shown in figure 3 is complicated because of the complex character of reduction process of W^{6+} oxide species, which proceeds through different intermediate oxidation states, depending on the reduction conditions, support, etc. However, it can be clearly seen that the relative proportion of different W oxide species and the temperature of their reduction were affected by the incorporation of Al, Ti and Zr oxides on the SBA-15 surface leading to a more homogeneous dispersion of deposited WO_3 species. In addition, some increase in the metal-support interaction after Al and Zr grafting on the SBA-15 was evidenced by an increase of hydrogen consumption at high temperature region (800-1000 °C). It seems that the addition of metal atoms (Al, Ti, Zr) on the SBA-15 surface prior to catalysts' preparation improve dispersion of W oxide species in calcined catalysts. Grafted metal species serve as anchoring sites for the heteropolyacid units providing a stronger interaction between the HSiW precursor and the supports. Results from the DRS characterization of NiW/(M)-SBA-15 catalysts (figure 4) are well in line with the above supposition. Thus, an increase in the absorption edge energy from 3.8 eV for the NiW/SBA-15 catalyst to 3.9-4.0 eV for the M-SBA-15-supported samples indicates a better dispersion of WO_3 species on Al, Ti, Zr-containing SBA-15 materials.

HRTEM characterization of sulfided catalysts showed that the dispersion of NiW active phase was also better on metal-containing SBA-15 supports than on the pure silica one (table II). The best dispersion of the active WS_2 phase was found for the catalyst supported on Zr-SBA-15.

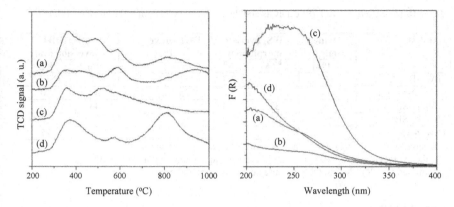

Figure 3. TPR profiles of NiW catalysts supported on: (a) SBA-15; (b) Al-SBA-15; (c) Ti-SBA-15 and (d) Zr-SBA-15.

Figure 4. UV-vis diffuse reflectance spectra of NiW catalysts supported on: (a) SBA-15; (b) Al-SBA-15; (c) Ti-SBA-15 and (d) Zr-SBA-15.

Catalytic activity

Catalytic activity of NiW/(M)-SBA-15 catalysts was evaluated in simultaneous hydrodesulfurization of DBT and 4,6-DMDBT. Figure 5 illustrates general reaction network for HDS of dibenzothiophenes. All NiW catalysts supported on metal-containing SBA-15 materials showed an outstanding catalytic performance in HDS of DBT and 4,6-DMDBT (table II). Their activity was more than 30 % higher in hydrodesulfurization of DBT and more than 3 times higher in HDS of 4,6-DMDBT than of the reference NiW/γ-alumina catalyst [10]. A good correlation was found between the dispersion of sulfided NiW active phase and catalytic activity results. The highest HDS activity was obtained with the NiW catalyst supported on Zr-containing SBA-15 molecular sieve, which makes it a promising catalytic system for ultra-deep HDS of diesel fuel (table II).

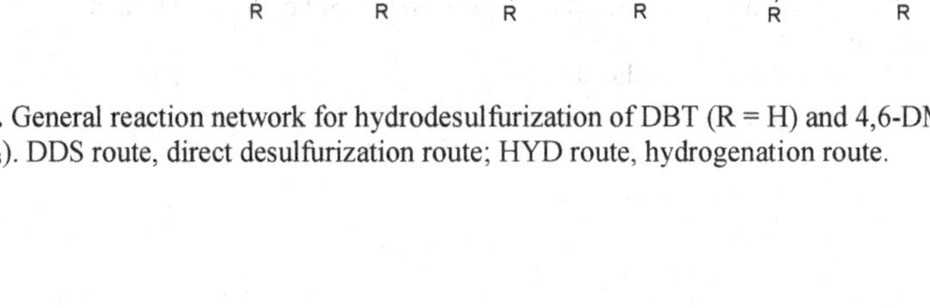

Figure 5. General reaction network for hydrodesulfurization of DBT (R = H) and 4,6-DMDBT (R = CH₃). DDS route, direct desulfurization route; HYD route, hydrogenation route.

Table II. Characteristics of NiW catalysts and their activity in HDS of DBT and 4,6-DMDBT.

Catalyst	Oxide catalysts α^a (%)	WS$_2$ average morphology L^b (Å)	N^b	DBT conversion (%) 4 h	8 h	4,6-DMDBT conversion (%) 4 h	8 h
NiW/SBA-15	64	41	2.3	51	74	29	47
NiW/Al-SBA-15	54	29	2.5	55	88	44	80
NiW/Ti-SBA-15	68	33	2.0	59	87	43	76
NiW/Zr-SBA-15	63	32	2.0	65	91	53	82
NiW/γ-Al$_2$O$_3$ [10]					65		27

[a] α, percentage of reduction of Ni and W oxide species determined by TPR.
[b] L, the average length of WS$_2$ slabs, and N, the average number of layers per WS$_2$ stack, determined by HRTEM from the measurement of at least 300 crystallites.

CONCLUSIONS

NiW catalysts supported on SBA-15-type materials modified with Al, Ti or Zr were prepared using HSiW heteropolyacid as W source. Characterization of the prepared catalysts showed high dispersion of W oxide and sulfided species on M-SBA-15 supports. All NiW/(M)-SBA-15 catalysts demonstrated an outstanding activity in HDS of DBT and 4,6-DMDBT.

ACKNOWLEDGMENTS

Financial support by CONACYT, Mexico (grant 100945) is gratefully acknowledged. The authors thank M. Aguilar Franco for technical assistance with small-angle XRD.

REFERENCES

1. A. Stanislaus, A. Marafi and M.S. Rana, *Catal. Today* **153**, 1 (2010).
2. T. Klimova, O. Gutiérrez, L. Lizama and J. Amezcua, *Micropor. Mesopor. Mater.* **133**, 91 (2010).
3. R. Palcheva, A. Spojakina, L. Dimitrov and K. Jiratova, *Micropor. Mesopor. Mater.* **122**, 128 (2009).
4. O.Y. Gutiérrez, G.A. Fuentes, C. Salcedo and T. Klimova, *Catal. Today* **116**, 485 (2006).
5. T. Klimova, J. Reyes, O. Gutiérrez and L. Lizama, *Appl. Catal. A: Gen.* **335**, 159 (2008).
6. L. Lizama and T. Klimova, *Appl. Catal. B.* **82**, 139 (2008).
7. A. Griboval, P. Blanchard, E. Payen, M. Fournier and J.L. Dubois, *Catal. Today* **45**, 277 (1998).
8. B. Pawelec, S. Damyanova, R. Mariscal, J.L.G. Fierro, I. Sobrados, J. Sanz and L. Petrov, *J. Catal.* **223**, 86 (2004).
9. D. Zhao, Q. Huo, J. Feng, B.F. Chmelka and G.D. Stucky, *J. Am. Chem. Soc.* **120**, 6024 (1998).
10. A. Soriano, P. Roquero and T. Klimova, *Stud. Surf. Sci. Catal.* **175**, 525 (2010).

Mater. Res. Soc. Symp. Proc. Vol. 1479 © 2012 Materials Research Society
DOI: 10.1557/opl.2012.1602

The Effect of Anodization Time on the Properties of TiO$_2$ Nanotube Humidity Sensors

Reshmi Raman, Oscar A. Jaramillo, and Marina E. Rincón*
CIE-UNAM, Priv. Xochicalco S/N, Col. Centro, Temixco, Mor. 62580.
*Corresponding author: merg@cie.unam.mx

ABSTRACT

In this paper, the effect of anodization time on the properties of TiO$_2$ nanotube humidity sensors is reported. TiO$_2$ nanotube arrays were grown by anodization of Ti foil using diethylene glycol and ammonium fluoride. Highly ordered TiO$_2$ nanotube arrays were obtained, with the length of tube increasing from 4 to 20 µm as the time of anodization increases, at the expense of nanotube integrity. Humidity sensors based on TiO$_2$ nanotube arrays were fabricated in impedance mode with ITO as top contact. The results revealed that sensor performance does not correlate with anodization time, regardless of enhanced area, showing an optimum morphology at 4h and 10h. The increase resistivity of the sensors upon water exposure, a donor molecule, is explained by the lack of TiO$_2$ stoichiometry and the fluctuations in the concentration of oxygen vacancies.
Keywords: sensor, nanostructures, morphology

INTRODUCTION

Humidity monitoring is very important for environmental control, industrial production, agriculture, food storage, and physicochemical processes in pharmaceuticals. In recent years, attention has been devoted to the development of high performance humidity sensors based on metal oxides [1–6], polymers [7–9], and inorganic/organic composite materials [10]. Humidity sensors require high sensitivity, low hysteresis, linear response, fast response and recovery behavior, wide humidity detection range, low cost, and excellent long-term stability [1-6, 11]. Of these sensors, the impedance type humidity sensors [3, 4, 11] are becoming more prevalent, in terms of quality and cost, than field effect transistors and fiber optic sensors. Furthermore, the thin film humidity sensors having nanosized grains and nanoporous structures have drawn much interest because of the high surface exposure for adsorption of water molecules. The adsorption of a water molecule on the surface of a metal oxide nanostructured semiconductor can be treated as a chemical–electronic process, in which charge transfer occurs between the adsorbed species and the dielectric matrix. This process can be greatly enhanced considering the low dimension and high surface-to-volume ratio of the nanostructured materials, which can cause a prominent change in the electrical conductivity or capacitance.

Due to the advantages of the high surface to volume ratio of many one-dimensional (1D) metal oxides, 1D-TiO$_2$ nanostructures such as nanofibers [12], nanowires [13], or nanotubes [4, 14], have also been synthesized for sensor applications. One-dimensional TiO$_2$ nanostructures, especially TiO$_2$ nanotubes arrays (TNA) formed by anodization of Ti foil, are structured by densely bundled nanotubes, which make the sponge-like films show large surface–volume ratio. They show excellent adsorptive ability for water vapor, which is critically required for fabrication of humidity sensors. This ordered nanoarchitecture offers more direct pathways for electrical transport and provides a higher possibility to form a surface electric field that should reduce the carrier recombination by confining carriers along the tube axis [4]. Moreover, the electrochemical anodization technique is capable of adjusting highly functional TiO_2 features (i.e., length, tube diameter, and self–ordering) over large length scales. Usually, long anodization time is correlated with longer nanotube lengths and improved device performance due to the enhanced area. In the present work we report the anodization of thin Ti foil by a conventional anodization path using fluorine based electrolyte. The results indicate that as humidity sensors, TNA respond more to surface reactivity than surface area, showing an anomalous response to water molecules in the humidity range studied.

EXPERIMENT

High purity Ti foil (Aldrich 99.99% pure) was thoroughly polished and ultrasonically cleaned with 50/50 ethanol-acetone mixture for 15 min. The samples were N_2 dried and loaded in a teflon cell containing 0.25 wt. % NH_4F and 1 wt. % H_2O in ethyleneglycole. Anodization was carried out at 50 V in a two-electrode cell under nitrogen atmosphere during 3 to 10 h. The as-deposited films were cleaned thoroughly in propanol and distilled water and subjected to air annealing at 550^0C for 3 h. Structural characterization was obtained using X ray diffraction (XRD, D/max 2200 V, Rigaku) with $CuK\alpha$ radiation source (λ=1.54 Å), and field emission scanning electron microscopy (FESEM, S-5500, Hitachi). Sensing measurements were performed as impedometric humidity sensors in a home built sensing set up (Figure 1). The impedometric sensors are based on impedance changes upon exposure to target species and were operated under alternating voltage of ± 10 mV in the frequency range of 10^{-2} to 10^6 Hz. Figure 1 shows details of the Ti/TNA/ITO sandwich type sensor loosely clamped to allow the flow of gas molecules through the active layer of the device. Water concentration was controlled by passing N_2 gas through a cold finger, using the methodology described in reference [15].

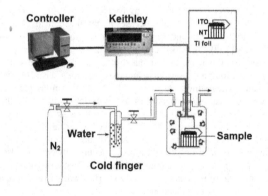

Figure 1. Schematic diagram of the humidity sensing set up

RESULTS AND DISCUSSION

Figure 2 shows the Bragg-Brentano (XRD) and grazing incidence (GXRD) patterns of TNA air annealed at 550^0C, where the peaks correspond to the anatase phase and the Ti substrate (around 2θ = 40°). With increasing anodization time, the intensity ratio between the (101) and (004) anatase planes becomes smaller, suggesting a major role of (004) planes at longer times.

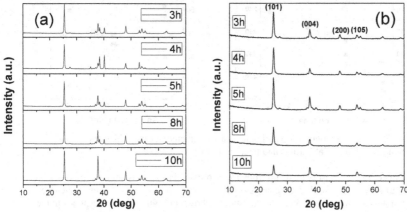

Figure 2. XRD patterns of Ti/TNT annealed samples as a function of anodization time in: (a) θ-2θ configuration; (b) grazing incidence of 0.5°.

Nanotube formation as a function of anodization time is shown in Figure 3. At 3 h the formation of nanotubes is not evident whereas at 4 h ring like structures join to form the tubes. At 5 h the tubes are well formed and spaced apart, and as time increases the tubes start to collapse. Further analysis of these images (Figure 4) indicates an increase in length and pore diameter with anodization time, and a reduction in wall thickness that leads to nanotube rupture.

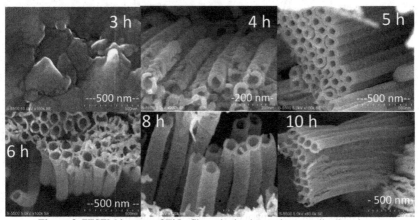

Figure 3. FESEM images of TiO$_2$ films obtained at various anodization times

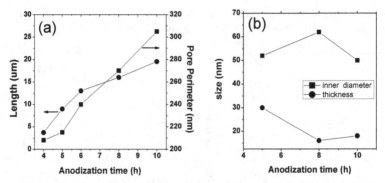

Figure 4. Evolution of length, pore perimeter, inner diameter and wall thickness of TiO$_2$ nanotube arrays obtained at various anodization times.

Impedometric sensors made up from TNAs obtained at various anodization times are expected to follow the morphology changes observed in Figures 3 and 4. A representative Nyquist plot as a function of water concentration is shown in Figure 5; it corresponds to the sensor response taken at room temperature using TNA anodized for 5h (5h-TNA) as the active film. The experimental curves are fitted using Z view software and the equivalent circuit (EQC) shown inside the figure. The EQC consisted of two nested loops composed of resistors and constant phase elements (CPE). Normally, the outer loop is for the high frequency region and describes fast processes occurring near the pore mouth (at the film/ITO interface). The inner loop is for the low frequency region and can be related to the capacitance inside the pores (near the Ti/film interface) and to charge transfer at inner defects or barrier oxide.

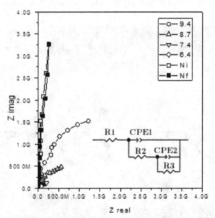

Figure 5. Nyquist plot of 5h-TNA impedometric sensor as a function of relative humidity (%RH from 6.4 to 9.4). Ni and Nf are the nitrogen baseline before and after water exposure.

Figure 6 compares the outer (R2) and inner (R3) resistors of the impedometric sensors as a function of water content and anodization time. For all TNAs, R2 and R3 show an increase in impedance as the concentration of water increases. For R2 (the external resistor not correlated with the interior of the pores), the lowest values correspond to the sample anodized for 4h. In contrast, R3 seems to follow the morphology change occurring at longer anodization time. It shows the largest values with 3h-TNA, but as the length of the tubes increases and the wall becomes thinner, R3 gradually decreases, except for 10h-TNA, where a sudden decrease is evident. The sudden decrease in resistance with anodization time can be attributed to an increase in surface reactivity due to the rupture of nanotubes wall. In the case of 4h-TNA, surface reactivity does not come from collapse since the nanotubes are just forming. In these samples, the nanotubes are ill-defined, relatively short, with a different ratio of crystalline planes, and without sharp inner and outer interfaces that cause more change in R2 than R3 upon water adsorption. Moreover, the value of CPE2 decreases at larger water content particularly for 4h-TNA, which shows the largest sensitivity (i.e., the largest capacitive change per water%). This is shown in Figure 7, where the capacitance of the TNA sensors increases at longer anodization time except for the 4h-TNA sample ($C_{4h} > C_{6h} > C_{5h} > C_{3h}$). The anomalous response to water (a donor molecule) of the resistive and capacitive elements suggest the lack of TiO_2 stoichiometry and abundance of surface traps and defects (i.e. oxygen vacancies, Ti^{3+} centers) in the air annealed TNAs, which can be responsible for the conduction type inversion (n-to-p), as has been reported for other 1-D TiO_2 nanostructures [16].

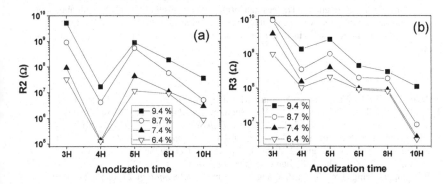

Figure 6. R2 (a) and R3 (b) as a function of humidity percentage and TNA anodization time.

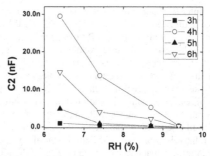

Figure 7. Capacitance of the CPE2 element as a function of RH% and TNA anodization time

CONCLUSIONS

The effect of anodization time on the properties of impedometric humidity sensors based on TiO$_2$ nanotube arrays was studied in detail. Highly ordered TiO$_2$ nanotube arrays were obtained, with the length of tube increasing from 4 to 20 μm as the time of anodization goes up, at expense of nanotube integrity. The sensing study revealed an anomalous sensor response to water, which can be explained due to the lack of TiO$_2$ stoichiometry and the fluctuation in the concentration of oxygen vacancies. The best performance was observed in TNAs with abundant surface defects and roughness, which can either be obtained at short (just forming) or long (breaking up) anodization time, indicating that surface reactivity is more important than effective surface area.

REFERENCES

1. Z. Wang, L. Shi, F. Wu, S. Yuan, Y. Zhao, M. Zhang, *Nanotechnology* **22**, 275502 (2011).
2. L.J. Yao, M.J. Zheng, H.B. Li, L. Ma, W.Z. Shen, *Nanotechnology* **20**, 395501 (2009).
3. L.Y. Li, Y.F. Dong, W.F. Jiang, H.F. Ji, X.J. Li, *Thin Solid Films* **517**, 948 (2008).
4. Q.Wang, Y.Z. Pan, S.S. Huang, S.T. Ren, P. Li, J.J. Li, *Nanotechnology* **22**, 025501 (2011).
5. H.N. Zhang, Z.Y. Li, W. Wang, C. Wang, L. Liu, *J. Am. Ceram. Soc.* **93**, 142 (2010).
6. N. Zhang, K. Yu, Z.Q. Zhu, D.S. Jiang, *Sensor Actuator B* **143**, 245 (2008).
7. X.F. Wang, B. Ding, J.Y. Yu, M.R. Wang, F.K. Pan, *Nanotechnology* **21**, 055502 (2010).
8. D.S. Han and M.S. Gong, *Macromol. Res.* **18**, 260 (2010).
9. X. Lv, Y. Li, P. Li, M.J. Yang, Sensor Actuator B **135**, 581 (2009).
10. C.H. Yuan, Y.T. Xu, Y.M. Deng, N. Jiang, N. He, L.Z. Dai, *Nanotechnology* **21**, 415501 (2010).
11. J. Wang, M.Y. Su, J.Q. Qi, L.Q. Chang, *Sensor Actuator B* **139**, 418 (2009).
12. J. Gong, Y.H. Li, Z.S. Hu, Z.Z. Zhou, Y.L. Deng, *J. Phys. Chem. C* **114**, 9970 (2010).
13. G. Wang, Q. Wang, W. Lu, J.H. Li, J. Phys. Chem. B **110**, 22029 (2006).
14. Y.Y. Zhang, W.Y. Fu, H.B. Yang, Q. Qi, Y. Zeng, T. Zhang, R.X. Ge, G.T. Zou, *Appl. Surf. Sci.* **254**, 5545 (2008).
15. J. A. Hatfield, *Environmental Progress* **23**, 45 (2004).
16. I.D. Kim, A. Rothschild, B.H. Lee, D.Y. Kim, S.M. Jo, H.L. Tuller, *Nano Lett.* **6**, (2009). (2006).

Mater. Res. Soc. Symp. Proc. Vol. 1479 © 2012 Materials Research Society
DOI: 10.1557/opl.2012.1603

Adhesive Energy of Zinc Oxide and Graphite, Molecular Dynamics and Atomic Force Microscopy Study

Ulises Galan,[1] Henry A. Sodano[2,3]

[1]School for Engineering of Matter, Transport and Energy, Arizona State University, USA
[2]Department of Mechanical and Aerospace Engineering, University of Florida, USA
[3]Department of Materials Science and Engineering, University of Florida, USA

Abstract

Molecular Dynamics (MD) simulations are performed to calculate the interfacial energy between zinc oxide (ZnO) and graphitic carbon for the study of solid–solid adhesion. The MD model consists of a ZnO slab and a single layer of graphitic carbon. The calculation was validated experimentally by atomic force microscopy (AFM) liftoff. A polishing process was applied to create a tip with a flat surface that was subsequently coated with a ZnO film allowing force displacement measurement on Highly Oriented Pyrolitic Graphite to validate the simulations. The MD simulation and AFM lift-off show good agreement with adhesive energies of 0.303 J/m^2 and 0.261 ± 0.054 J/m^2, respectively.

Keywords: Molecular Dynamics, Atomic Force Microscopy, Interface Strength

Corresponding author. mgalanve@asu.edu

Introduction

Recently Lin at el [1] employed zinc oxide (ZnO) nanowires (NWs) as an interphase between the carbon fibers and polymer matrix in fiber reinforced composites and demonstrated a more than two times increase in the interfacial shear strength compared to bare fibers. The research was extended to investigate the properties of ZnO NWs on aramid fibers [2] as well as to study the role of morphology in the interfacial, [3] ultimately achieving 3.27 times the interface strength offered by a composites grade epoxy. The results suggested that the adhesion of the ZnO NWs with the fibers was responsible for the improved interfacial shear strength. In this context, the study of the surface interactions between ZnO and carbon offers the potential for further enhancement to this system as well as the general problem of adhesion to graphitic surfaces. With the aim to better understand the mechanical properties between ZnO and carbon, we develop a MD model and validate the model with AFM lift off measurements from the surface of Highly Oriented Pyrolitic Graphite (HOPG) using a ZnO coated AFM tip.

Experiments

MD simulates the atomic interactions that govern the energetics and properties of a material by describing these interactions with a force field and assuming atoms as single points with a defined mass [4]. A range of material properties such as adhesion, diffusion, cracking, creep, etc. can be studied by applying the laws of classical mechanics, thus MD provides a tool to study systems at an atomic level. The attractive and repulsive forces that arise at the atomic scale are defined by a force field can be divided into bonded and non-bonded interactions for neutral systems, while ionic interactions are accounted for by the summation of the electrostatic energies. The ionic interactions in ZnO can be described by pairwise interactions with the

Buckingham [5] force field and the charges of the ions can be described through the Ewald method [6]. In the case of graphitic carbon, the optimized potential for liquid simulations (OPLS) [7] is used, which includes bonded, non-bonded, angular and torsional parameters. The Zn-Carbon and O-Carbon parameters were taken from reference [8] and [9] respectively, and apply to the Lennard-Jones potential which is used to describe non bonded interactions. For the MD model, the large-scale atomic molecular massively parallel simulator (LAMMPS) [10] has been used to perform the simulation. The MD model of the interface consists of ZnO with free surfaces (0001), $(000\bar{1})$, and a single layer of graphite (graphene).

ZnO is an ionic solid with wurtzite structure and lattice parameters of a = 3.25 and c = 5.20. The structure consists of alternating planes of coordinated O^{2-} and Zn^{2+} ions along the c axis producing a dipole moment along this axis that gives rise to polar surfaces. The polar surfaces are Tasker type III [11] either Zn (0001) or O $(000\bar{1})$ terminated and the structure generates an electrostatic field perpendicular to the surfaces, the accumulation of electrostatic energy causes the surface energy to diverge and instability of the structure. These surfaces have been discussed in the literature [11--14] and it has been found that the surface morphologies are dependent on the polarity of the surfaces. One way to stabilize the structure is by removing one fourth of Zn atoms from the (0001) surface [15]. Still further research has shown that the (0001) surface undertakes triangular shape reconstruction in the stabilization process [12]. For the MD model used here, ZnO was constructed with periodic boundary conditions in the X and Y directions to give stability to the relaxed structure. The Z direction is not periodic and was stabilized considering two cases of the polar surfaces. First, one fourth of Zn atoms were removed from the Zn terminated surface; while in the second case, triangular voids were formed at the (0001) surface [12]. The edges of the holes have oxygen atoms and the dimension of the edge was 9.3 Å. In both cases a stable configuration was achieved. The dimensions of the structure were 42 Å by 38 Å by 20Å. In the first case with one fourth of the Zn atoms removed, the ZnO model has a total of 2912 atoms and in the second case with triangular voids a total of 2765 atoms. ZnO has multiple nonpolar surfaces, in these surfaces the energy is finite and it is possible to have stable free surfaces without distortion of the ideal wurtzite crystal structure. With the aim to capture the main properties of the ZnO crystals deposited on the AFM tip here we also performed a simulation on the $(1\bar{1}00)$ nonpolar surface, with the model having 2912 atoms and only relaxation of the structure being necessary. The isothermal-isobaric ensemble [16] (NPT) was used to obtain zero stress at the X and Y boundaries of ZnO. Specifically, the temperature was set to 100 K and the system was simulated for 20 ps with the temperature gradually reduced to 1 K in the interval of 20 ps while keeping the NPT ensemble. The canonical ensemble (NVT) [17,18] was used to keep the volume constant and temperature of about 1 K for 20 ps.

After relaxation, the NVT ensemble was used to keep the temperature constant and a displacement was applied to the graphene to simulate liftoff. The positions of the atoms at the top of ZnO were fixed in space; graphene was separated in the negative Z direction while the rest of the atoms of ZnO were allowed to deform through the interaction forces with the graphene layer. During the simulation, the stress on the graphene was recorded. The layer of graphene was displaced by steps of 0.25 Å and the structure was allowed to equilibrate for 20 ps in between each step, to yield a test velocity of 1.25 m/s. The stress on graphene during the liftoff is shown in Figure 1. The first step of separation generated an adhesive force of 9.8 nN which rose rapidly in the subsequent steps to achieve a maximum adhesive force of 14.7 nN at 0.75 Å of separation. The force then begins a more gradual decline and is approximately zero nN at 7

Å. For the polar surface, both ZnO stabilization cases, one fourth of Zn atoms removal and triangular void stabilization were considered in the simulated separation and the results obtained were the same. For the (1100) nonpolar surface the separation simulation was performed with nearly the same results as the polar surface. The interface energy was calculated by integrating the force displacement curve to yield a total liftoff energy of 2.734 aJ. Normalizing by the area of the graphene surface (896 Å^2) gives a specific liftoff energy of 0.303 J/m^2.

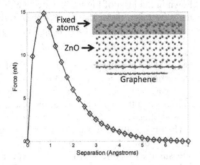

Figure 1. Force displacement curve on carbon atoms for the separation between the two materials as calculated with LAMMPS.

The silicon tip was flattened via a polishing over a diamond lapping film with a grain size of 0.1 μm to create a smooth surface and one that accounts for the slope in the cantilever such that the flattened tip is parallel to the substrate surface. Once polished the surface area of the tip was measured in a scanning electron microscope to be 0.067 □m^2, as shown in Figure 2. The uncoated polished tip was repeatedly tested with HOPG and the interface energy was measured to be 0.087 J/m^2 with a standard deviation of 0.0071 J/m^2, as shown in Figure 3 (blue squares).

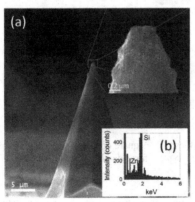

Figure 2. a) Flat surface of AFM tip after polishing and b) EDX showing Zn.

The thin, conformal ZnO film was created through the deposition and coalescence of ZnO nanoparticles. The ZnO nanoparticles were synthesized following the methods of Hu et al [19] to obtain a stable colloidal suspension of nanoparticles. Once the solution was ready the AFM tip was dipped into the solution and then put into an oven at 70 °C for 3 minutes. This process was repeated seven times to form a thin layer of ZnO on the tip. Energy dispersive x-ray spectroscopy in a scanning electron microscope confirmed the formation of the ZnO layer on the surface of the AFM tip and is shown in the inset of Figure 2. Once the tip was coated with ZnO, force displacement measurements were taken under the AFM with a fresh HOPG surface. The AFM tip approaches the surface of HOPG and a small force is applied that deflects the cantilever to a load of 40 nN.

The presence of ZnO on the AFM tip produces a higher energy of adhesion as shown in Figure 3. As a calibration of the testing procedure, the ZnO layer was then removed using an acid solution (HCl) and the testing was repeated, producing the same adhesive energy prior to coating with ZnO. This both demonstrated the repeatability of the test as well as the validity of the approach. A layer of gold was sputter coated on top to serve as a reflective surface that is unaffected by the acid cleaning. The tip was then recoated with ZnO using the same process and the interface energy was measured followed by cleaning of the tip in acid and testing on a cleared HOPG surface. The measurements were confirmed by repeating this process three additional times, with the mean for each set of tests shown in Figure 3. The average value for the energy of adhesion was 0.261 J/m^2 with a standard deviation of 0.054 J/m^2 for the ZnO HOPG interface.

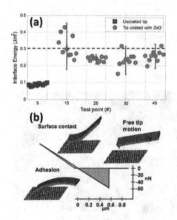

Figure 3. a) Results of AFM test. Blue square uncoated tip, red point tip coated with ZnO. b) Typical force displacement curve

Discussion

The liftoff testing demonstrates the strong adhesive energy between ZnO and HOPG. Furthermore, the results from the simulation and the AFM liftoff test show good agreement. It should however be noted that the nature of the experiment is somewhat different from the simulation. Specifically, in the AFM liftoff the cantilever deflects and the two surfaces separate because of the potential energy accumulated in the cantilever beam. In the case of MD the separation was applied along the Z direction, graphene was assumed to move rigidly and the reaction forces are recorded by LAMMPS and integrated to measure the adhesive energy. The calculation of the energy of adhesion between ZnO and HOPG was computed through the force displacement curves of a liftoff experiment. In both the MD simulation and the AFM liftoff experiment, the adhesive energy is similar, with the MD model yielding 0.303 J/m^2 and the AFM experiments yielding 0.261 ± 0.054 J/m^2. The nature of the AFM liftoff experiment measures the adhesion energy by computing the area formed by the shaded triangle as shown in Figure 3. The force measured at the moment of separation in the AFM was 66.53 ± 7.46 nN. The MD simulation recorded a maximum force of 14.7 nN. This difference stems from the nature of the experimental setups because the AFM cantilever beam stores all of the potential energy needed to separate the surfaces before the interface fails with an abrupt change in force. The MD simulation of the interface computes the force as the graphene and the ZnO separate, a process that is not possible to do experimentally. This can be seen in Figure 1, the force reaches the maximum and decays slowly because the interaction forces and more generally the force field used to represent the interaction of each atomic species across the interface, whereas in AFM measurement and the force displacement curve shows a triangular area that decays sharply once the tips separates form HOPG. Since the shape of the force displacement curves is different, the same maximum force is not expected. However in both cases the force-displacement curve is used to calculate the adhesive energy per surface area. Taking all of these factors into account, the MD model was utilized to predict the energy of adhesion between ZnO and HOPG surfaces and AFM liftoff experiments validated the calculations to be accurate.

Conclusion

This letter has investigated the adhesion between ZnO and graphite, which was recently shown to provide 3.28 times higher shear strength than epoxy in fiber reinforced composites [3]. The adhesion was simulated using MD and shown to have an adhesive energy of 0.303 J/m^2 which is more than 4.3 times higher than predicted between carbon nanotubes and epoxy [20]. The strong boding present between ZnO and carbon may lead to methods for the enhancement of organic interfaces through the use of inorganic interphase. Additionally, the simulations were validated through AFM lift off studies that showed very good agreement with the models. This validation approach can be used to further enhance the impact of numerical predictions of properties in solid-solid interfaces.

Acknowledgments

The authors would like to acknowledge the Army Research Office (Award # W911NF0810382) and Dr. David Stepp for funding of this project.

[1] Y. Lin, Advanced functional materials 19, 2654 (2009).

[2] G. J. Ehlert, ACS applied materials & interfaces 1, 1827 (2009).

[3] U. Galan, Composites Sci. Technol. (2011).

[4] J. P. Ryckaert, Journal of computational physics 23, 327 (1977).

[5] D. Binks, *Computational Modeling of Zinc Oxideand related oxide ceramics* (University of Surrey, Harwell, England, 1994).

[6] T. Darden, J. Chem. Phys. 98, 10089 (1993).

[7] W. L. Jorgensen, J. Am. Chem. Soc. 118, 11225 (1996).

[8] J. Y. Guo, J. Appl. Phys. 109, 024307 (2011).

[9] J. G. Harris, Journal of physical chemistry (1952) 99, 12021 (1995).

[10] S. Plimpton, Journal of computational physics 117, 1 (1995).

[11] P. W. Tasker, Journal of physics.C, Solid state physics 12, 4977 (1979).

[12] O. Dulub, Phys. Rev. Lett. 90, 16102 (2003).

[13] G. Kresse, Physical review.B, Condensed matter 68, 245409 (2003).

[14] Y. M. Yu, Physical review.B, Condensed matter and materials physics 77, 195327 (2008).

[15] G. Heiland, Surf. Sci. 13, 72 (1969).

[16] H. J. C. Berendsen, J. Chem. Phys. 81, 3684 (1984).

[17] S. Nosé, Mol. Phys. 50, 1055 (1983).

[18] H. C. Andersen, J. Chem. Phys. 72, 2384 (1980).

[19] Z. Hu, J. Colloid Interface Sci. 263, 454 (2003).

[20] J. Gou, Computational materials science 31, 225 (2004).

Mater. Res. Soc. Symp. Proc. Vol. 1479 © 2012 Materials Research Society
DOI: 10.1557/opl.2012.1604

Effect of the Nucleation Layer on TiO$_2$ Nanoflowers Growth via Solvothermal Synthesis

Oscar A. Jaramillo, Reshmi Raman, and Marina E. Rincón*
CIE-UNAM, Priv. Xochicalco S/N, Col. Centro, Temixco, Mor. 62580, México.
*Contact email: merg@cie.unam.mx

ABSTRACT

TiO$_2$ nanoflowers were obtained on modified ITO substrates by solvothermal synthesis. Surface modification was achieved with a layer of TiO$_2$ seeds/nucleus obtained by dip-coating at various pH and dip cycles. Field emission scanning electron microscopy results indicated that at all nucleation conditions there was a dual population of TiO$_2$ nanoparticles and nanoflowers. For a particular pH, the effect of increasing the number of dips was to increase the size and number of the nanoflowers, whereas for a fixed number of dips, the increase in pH causes a decrease in nanoflower population. The comparison with solvothermal films obtained on unmodified substrates indicates that TiO$_2$ nanoflowers grew up on the nucleation sites. These microstructural changes determine the active surface area and sensing properties of the solvothermal films. At room temperature, no evidence of superior ethanol sensing properties was found for TiO$_2$ nanoflowers, which show larger resistivity than TiO$_2$ nanoparticles.
Keywords: crystal growth, nanostructure, sensor

INTRODUCTION

Over the past decades, nanostructured materials have emerged as promising materials for fundamental studies and possible technological applications. The reduction in size of functional architectures has been a dominating trend in many fields of science and technology. This size reduction provides unique physical and chemical properties, such as quantum confinement and high electron mobility. In particular, TiO$_2$ nanostructures have received a great deal of attention owing to its outstanding chemical and physical properties and have been extensively used in a wide range of environmental and energy related applications [1-5]. For many of these applications, including gas sensors, it is important to maximize not only the specific surface area of TiO$_2$ nanostructures to increase the device performance, but also to take advantage of new effects associated with particular geometries, such as those found in TiO$_2$ nanotubes and nanowires related to surface reconstruction and surface curvature. Consequently, TiO$_2$ nanoparticles and one-dimensional TiO$_2$ thin films have been widely used as gas sensors in the detection of acetone, ethanol, hydrogen, and carbon monoxide [6-9].

Synthesis of TiO$_2$ nanostructures with confined geometry may be achieved by electrochemical anodic oxidation, sol-gel, chemical vapour deposition, and hydrothermal or solvothermal synthesis [10-15]. Among these methods, hydrothermal/solvothermal synthesis of TiO$_2$ is a promising approach due to its simplicity, fast reaction velocity, and low cost. The advantages of the non-hydrolytic synthesis lie in the suppression of uncontrolled hydrolysis at the beginning of the condensation reaction [15]. Therefore, the solvothermal method has the potential for depositing novel TiO$_2$ nanostructures as thin films. So far, the typical geometries obtained by solvothermal synthesis on transparent conductive oxide substrates have been nanorods, nanowires, and nanoribbons [12-14], while TiO$_2$ nanotubes with small aspect ratio have been obtained as powders using variants of the hydrothermal method reported by Kasuga et. al. [16-18].

In the present work, TiO_2 nanoflowers were obtained on modified ITO substrates by solvothermal synthesis. Surface modification was achieved with a layer of TiO_2 seeds /nucleus obtained by dip-coating at various pH and dip cycles. The comparison with solvothermal films obtained on unmodified substrates indicates that TiO_2 nanoflowers grew up on the nucleation sites, and its number and size depend on the characteristics of the dip-coating bath. The ethanol sensing properties of solvothermal films compose of nanoparticles and nanoflowers were compared to evaluate the sensing properties of the nanoflower structure.

EXPERIMENT

Indium Tin Oxide (ITO) substrates were cleaned thoroughly with distilled water, 2-propyl alcohol, acetone, and ethanol, successively in an ultrasonic bath for 10 minutes. The substrates were air dried before being dip-coated in a sol gel bath to provide the nucleation layer for solvothermal synthesis. Titanium tetraisopropoxide (TTIP) was used as the titanium precursor for both the nucleation layer and the solvothermal film. At pH 0, the sol-gel bath contains 8 mL of TTIP (Sigma Aldrich 97%), 84 mL of 2-propanol (Sigma Aldrich 99%), and 8 mL of HCl (J.T. Baker 37 wt. %). The mixture of 2-propanol and HCl was kept under strong stirring and TTIP added drop wise. The entire solution was stirred for 2 h and aged for 24 h. The pH of the dipping solution was adjusted by changing the amount of acid, whereas the pH of the solvothermal solution was always kept at pH 0. The substrates were dipped into the sol gel solution and withdrawn at a constant rate of 30 mm/min with a delay of 10 s. The samples were placed in a Teflon liner inside an autoclave operated at 150°C for 6 h. After the solvothermal reaction, all the substrates were washed thoroughly in distilled water and isopropyl alcohol, dried in air, and subjected to air annealing at 550°C for 3h. Samples designated as D1, D3, and D5 refer to those obtained on ITO substrates modified with a nucleation layer obtained at a fixed pH and 1, 3, and 5 dips, respectively. Those designated as P0, P2, and P6, were obtained at a fixed number of dips (5) and pH 0, 2, and 6, respectively. Solvothermal films obtained on clean ITO substrates were labeled S.

The structural characterization of the samples was carried out by X ray diffraction (XRD, D/max 2200 V, Rigaku) with CuKα radiation source (λ=1.54 Å). The microstructure of the samples was investigated using field emission scanning electron microscopy (FESEM, S-5500, Hitachi). Sensing experiments were carried out in a home built system (Figure 1), with nitrogen as the carrier gas. The vol. % of ethanol in the sensing chamber was calculated according to the methodology described in reference [19]. The sample electrical connections were made by two parallel silver lines separated by equal distance (5 x 5 mm^2), it allows for sheet resistance measurements under N_2 flow (R_N) and after exposure to some vol. % ethanol (R_{eth}). The current-voltage characteristics were measured using a picoammeter/voltage source measure unit (6487, Keithley). The resistor platform was selected for the sensing studies due to the low conductivity of the ITO substrate after exposure to the solvothermal conditions. The sensor response S was defined as the relative change of sheet resistance:

$$SR = (R_N - R_{eth})/R_{eth} \times 100\% \qquad (1)$$

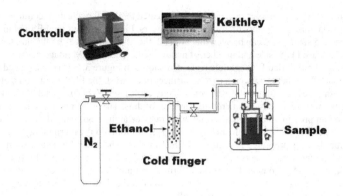

Figure 1. Schematic diagram of the ethanol sensing set up

DISCUSSION

Figure 2(a) shows the XRD patterns of solvothermal films grown on nucleation layers formed by varying the number of dips. All as-deposited films were found to be amorphous and became crystalline after annealing at 550°C for 3h. Without the nucleation layer, the solvothermal film shows a small intensity peak assigned to the (101) anatase plane. On titania seeds the rutile phase becomes dominant, particularly at increasing number of dips. Figure 2(b) shows the effect of decreasing the acidity of the dipping bath on the crystallinity of the solvothermal films; here the lower intensity of the (110) rutile plane suggests thinner films as the pH increases. Notice the absence of ITO diffractions peaks, since it undergoes dissolution under the conditions of the solvothermal synthesis.

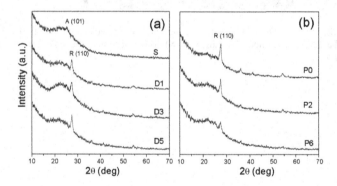

Figure 2. XRD patterns of solvothermal films as a function of: (a) number of dips at pH 2; (b) pH of the dipping solution at 5 dips.

Analysis of the samples microstructure indicated that at all nucleation conditions there was a dual population of nanoparticles and nanoflowers. Figure 3 shows the FESEM images of D1, D3, and D5 samples. It can be seen that at pH 2, the effect of increasing the number of dips causes an increase in the size and number of nanoflowers. The FESEM image of the olvothermal film grown on unmodified ITO (sample S) is also shown in Figure 3. S is characterized by small TiO_2 nanoparticles (anatase) given a porous surface. In contrast, the films grown on a nucleation layer show the presence of nanoflowers (most likely rutile), which become more dominant as the seed density increases due to the increasing number of dips. These microstructural changes are the result of suppressing the random formation of nucleus at the beginning of the condensation reaction. The effect of changing the pH of the dipping solution on the morphology of the solvothermal films is presented in Figure 4. FESEM images show that for a nucleation layer formed with 5 dips, nanoflowers population decreases with the increase in alkalinity of the dipping bath. At pH 6, no nanoflowers were formed at 1 and 3 dips. Apparently, anatase nanoparticles dominate in the absence of a nucleation layer and the pH of the dipping solution plays an important role on the growth of nanoflowers.

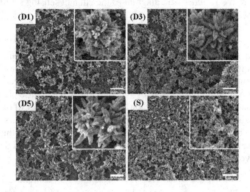

Figure 3. FESEM images (D1, D3, D5) of solvothermal films grown on modified ITO substrates, and also on unmodified ITO substrates (S). Magnification of the main features are shown in the inset.

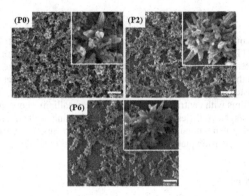

Figure 4. FESEM images of solvothermal films grown on ITO substrates modified by a nucleation layer obtained after 5 dips at pH 0 (P0), 2 (P2), and 6 (P6). Magnification of the main features are shown in the inset.

The optical characterization of the solvothermal films (not shown) revealed that they are highly transparent (80 % transparent in the visible region of the spectrum). To determine the ethanol sensing properties of nanoflowers, the SR of solvothermal films dominated by nanoparticles (S) was compared to that of solvothermal films dominated by nanoflowers (D5). Figure 5 shows that in both films there is a clear increase in SR due to the decrease in sheet resistance of the solvothermal films after exposure to ethanol. The sensing mechanism can be attributed to target vapour molecules diffusing into the sensing layer through the pores and displacing remaining adsorbed oxygen (an electron sequester) and even donating lonely electrons, which increases the conductivity of n-type TiO_2 [7]. The fact that D5 has an inferior sensor response and sensitivity (i.e., the sensitivity is defined as the slope of Figure 4), could either be due to the dominance of a different crystalline phase (Rutile vs. Anatase), and/or to an inferior surface area.

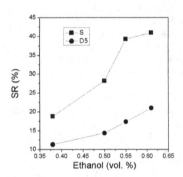

Figure 5. Sensor response (SR) of solvothermal TiO_2 films: S (nanoparticles), D5 (nanoflowers).

CONCLUSIONS

Highly transparent TiO_2 films were synthesized on modified ITO substrates by solvothermal synthesis. ITO modification was accomplished by dip-coating a thin layer of TiO_2 seeds. The effect of increasing the number of dips causes an increase in the size and number of nanoflowers whereas both decrease with an increase in pH. Without the nucleation layer, the solvothermal film consisted on a porous nanoparticulated anatase film. Ethanol TiO_2 sensors based on solvothermal films with contrasting geometry and crystallinity show similar tendencies in both transparent sensors. Nevertheless, the larger surface area and/or the dominance of anatase on the solvothermal film grown without a nucleation layer, impart superior sensing properties in the resistor platform. Other sensing platforms could be more suitable for the nanoflower geometry.

ACKNOWLEDGMENTS

To M. Sánchez, M. Solís, R. Morán, and M. L. Ramón for technical support, and to projects CONACYT 153270 and DGAPA-UNAM IN106912 for financial support.

REFERENCES

1. P. Roy, S. Berger and P. Schmuki, *Angew. Chem. Int.* **50**, 2904 (2011).
2. Y. Sun, S. Liu, F. Meng, J. Liu, Z. Jin, L. Kong and J. H. Liu, *Sensors* **12**, 2610 (2012).
3. T. Djenizian, I. Hanzu and P. Knauth, *J. Mater. Chem.* **21**, 9925 (2011).
4. C. Chang, C. Chen, C. Chen, and C. Kuo, *Thin Solid Films* **520**, 1546 (2011).
5. F. Shao, J. Sun, L. Gao, S. Yang and J. Luo, *J. Phys. Chem. C* **115**, 1819 (2011).
6. R. Rella, J. Spadavecchia, M. Manera, S. Capone, A. Taurino and M. Martino, *Sens. Acuators B* **127**, 426 (2007).
7. Y. Wang, S. Tan, J. Wang, Z. Tan, Q. Wu, Z. Jiao and Q. Wu, *Chin. Chem. Lett.* **22**, 603 (2011).
8. M. Paulose, O. Varghese, G. Mor, C. Grimes and K. Ong, *Nanotechnology* **17**, 398 (2006).
9. Z. Seeley, A. Bandyopadhyay and S. Bose, *Thin Solid Films* **519**, 434 (2010).
10. H. Omidvar, S. Goodarzi, A. Seif and A. R. Azadmehr, *Superlattice Microst.* **50**, 26 (2011).
11. M. Addamo, V. Augugliaro, A. Di Paola, E. Garca-lopez, V. Loddo, G. Marc and L. Palmisano, *Thin Solid Films* **516**, 3802 (2008).
12. B. Liu and E. Aydil, *J. Am. Chem. Soc.* **131**, 3985 (2009).
13. Z. Wei, L. Roushi, T. Huang and A. Yu, *Electrochim. Acta* **56** 7696 (2011).
14. T. Beuvier, M .R. Plouet, M. M. Le Granvalet, T. Brousse, O. Crosnier and L. Brohan, *Inorg. Chem.* **49**, 8457 (2010).
15. X. Wang and P. Xiao, *J. Mater. Res.* **21**, 1189 (2006).
16. W. C. Lin, R. Liu, W. D. Yang, Z. J. Chung and H. J. Chueng, *Adv. Mat. Res.* **391-392**, 1334 (2012).
17. Y. Li, X. Y. Yang, Y. Feng, Z. Y. Yuan and B. L. Su, *CRC Cr. Rev. Sol. State* **37**, 1 (2012).
18. T. Kasuga, M. Hiramatsu, A. Hoson, T. Sekino and K. Niihara, *Langmuir* **14**, 3160 (1998).
19. J. A. Hatfield, *Environmental Progress* **23**, 45 (2004).
20. P. Hu, G. Du, W. Zhou, J. Cui, J. Lin, H. Liu, D. Liu, J. Wang and S. Chen, *J. Am. Chem.* **2**, 3263 (2010).

Mater. Res. Soc. Symp. Proc. Vol. 1479 © 2012 Materials Research Society
DOI: 10.1557/opl.2012.1605

Thermal Conductivity of Composites with Carbon Nanotubes: Theory and Experiment

J. Ordonez-Miranda[1], C. Vales-Pinzon[1], J. J. Alvarado-Gil[1]
[1]Department of Applied Physics, Cinvestav, Carretera Antigua a Progreso km. 6, A.P. 73
Cordemex, Merida, Yucatan, 97310, Mexico.
E-mail: eordonez@mda.cinvestav.mx

ABSTRACT

In this work, the thermal conductivity of composites made up of carbon nanotubes embedded in a polyester resin is investigated by comparing experimental data with theoretical predictions. The composite samples were prepared with a random and aligned distribution of carbon nanotubes. Its thermal conductivity is then measured by using the photothermal radiometry technique. The obtained experimental data is accurately described by the proposed theoretical model, which takes into account the size effects of the nanotubes. It is expected that the obtained results can provide useful insights on the thermal design of composites based on carbon nanotubes.

Keywords: Composite, Nanostructure, Thermal conductivity.

INTRODUCTION

Over the past few decades, significant research efforts have been devoted to the study of the thermal properties of particulate composites, due to their many technological applications ranging from mechanical structures to electronics[1, 2]. Carbon materials, as carbon nanotubes, nanoplatelets, graphene oxide nanoparticles and graphene flakes; are among the most promising filler particles to maximize the enhancement of the usually low thermal conductivity of the matrix [1]. For instance, enhancements above 100% have been reported for a small concentration of 1% of carbon nanotubes or graphene loading,[1] which have a good coupling to the matrix materials and a geometry that favors the heat conduction through them. Despite their importance, the thermal performance of these smart materials is not well-understood to date, especially at high volume fractions of micro/nano-sized particles.

In this work, the thermal conductivity of various composites with carbon nanotubes embedded in a polyester resin is measured by using the photothermal radiometry technique[3]. The composite samples were prepared with a random and aligned distribution of carbon nanotubes (CNT), which have a diameter ranging from 50nm to 80nm. The predictions of the proposed approach are in good agreement not only with the experimental data recorded in our laboratory, but also with the one reported in the literature [1]. It is expected that the obtained results can provide useful insights on the thermal design of composites based on carbon nanotubes.

THEORY

A suitable model to describe the thermal conductivity of composites made up with low concentrations of spheroidal particles embedded in a matrix has been proposed by Nan et al[4]. For cylindrical particles, this model reduces to

$$\frac{k_{\parallel}}{K_m} = 1 + \left(K_{pm} - 1\right)f \qquad (1a)$$

$$\frac{k_{random}}{K_m} = \frac{1 + Bf}{1 - 2Af} \qquad (1b)$$

where K_m and K_p are the bulk thermal conductivities of the matrix and particles, respectively; $K_{pm} = K_p / K_m$, f is the volume fractions of the particles, and

$$A = \frac{1}{3}\frac{K_{pm} - 1}{K_{pm} + 1}, \qquad (2a)$$

$$B = A\left(K_{pm} + 3\right), \qquad (2b)$$

When the cylindrical particles are aligned $k_{\parallel}$ determines the thermal conductivity of the composite along the axis of the cylinders and k_{random} is its thermal conductivity when the cylinders are randomly distributed.

Given that Nan et al model[4] is based on the Fourier law of heat conduction Eqs.(1) and (2) are expected to be suitable in predicting the thermal conductivity of composites with macro- and micro-sized particles. However, their validity is questionable for nanocomposites, where the particles size is of the order of or even smaller than the mean free path (MFP) of the energy carriers (electrons and phonons), as is the case of the CNT. This is because the matrix and particle thermal conductivities in nanocomposites are not equal to their bulk values due to increased interface scattering of the energy carriers.

By taking into account the scattering of the energy carriers inside the matrix with the surface of the particles, Ordonez-Miranda et al.[5] have shown that the thermal conductivity k_m of the matrix is determined by

$$k_m = \frac{K_m}{1 + l_m \sigma_{\perp} f}, \qquad (3)$$

where l_m is the bulk MFP of the energy carriers inside the matrix and in absence of the particles, and the collision cross section per unit volume $\sigma_{\perp} = A_{\perp}/V$, being V the volume of one particle and $A_{\perp}$ the average collision cross-section (the effective area of collision) between an energy carrier and a particle. Equation (3) indicates that due to the presence of the particles inside the matrix, its bulk thermal conductivity is reduced by the factor $(1 + l_m \sigma_{\perp} f)^{-1}$, which involves the relative size of the particles with respect to the MFP l_m and the volume fraction f. Thus, the effect of the carrier-particle scattering increases as the ratio collision cross-section/volume increases. For spherical particles of radius a, the energy carriers of the matrix "see" an effective area of collision $A_{\perp} = \pi a^2$, which implies that $\sigma_{\perp} = \pi a^2 / \left(4\pi a^3/3\right) = 3/4a$. On the other hand, for cylindrical particles of radius a and length L, $A_{\perp} = 2aL$, and therefore $\sigma_{\perp} = 2aL / \left(\pi a^2 L\right) = 2/\pi a$.

On the other hand, by considering the scattering of the energy carriers inside the particles with their surface, Ordonez-Miranda *et al.*[5] have shown that the thermal conductivity k_p of nanoparticles is given by

$$k_p = \frac{K_p}{1 + l_p/c}, \tag{4}$$

where K_p is the bulk thermal conductivity of the particles and c is the average distance traveled by the energy carriers due to their scattering with the boundary of the particles and independently of the intrinsic carrier scattering, which is represented by the MFP l_p. Equation (4) shows that when the characteristic length c of the particles is much larger than the bulk MFP ($c \gg l_p$), the effective thermal conductivity reduces to its macroscopic value, as expected. However, for $c \le l_p$ the effective thermal conductivity can be considerably smaller than its bulk value. According to Ordonez-Miranda *et al.* results [5], for spherical and cylindrical particles of radius a, $c = a$ and $c = 32a/3\pi^2$; respectively.

Note that according to Eqs.(3) and (4), the reduction of the thermal conductivity of the matrix (particles) with respect to its bulk value increases with the ratio l_m/a (l_p/a) between the mean free path of the energy carriers and the size of the particles. For large particles sizes ($a \gg l_m, l_p$), both thermal conductivities of the matrix and particles reduces to their corresponding bulk values, as expected.

EXPERIMENT

Sample preparation

The studied composite material has been elaborated by embedding multi-walled CNTs in a non-polymerized polyester resin (RESINMEX MR-227), which is in liquid state. The CNTs have a diameter of 50-80 nm, a length of 0.5-2 µm and they were manufactured by Nanostructured & Amorphous Materials Inc. The obtained mixture was gently stirred for 10 min to generate a uniform mixing. A catalyzer of black ink was then added to induce polymerization and the composite was introduced in a cylindrical silicone mold of 10 mm of diameter and 3 cm of height. A ferrofluid (Ferrotec Corporation) with magnetite particles (Fe_3O_4) of an average diameter of 10nm [6], was used to prepare the composite samples with CNT volume fractions of 0.1%, 0.2%, 0.5%, 1%, 2% and 4%. The magnetization of the CNTs was performed through the loading of the magnetite nanoparticles [7, 8]. The samples were then put during 15 minutes inside of a uniform magnetic field of 600 G generated by a pair of Helmholtz coils, as illustrated by Medina-Esquivel et al[9]. The magnetic field induced the alignment of the CNTs inside the matrix and kept their arrangement due to the polymerization process, which after 24 hours produces hard samples. The obtained cylindrical sample was cut in circular plates of around 1 mm of thickness and polished carefully with sandpaper up to a thickness of 300 µm.

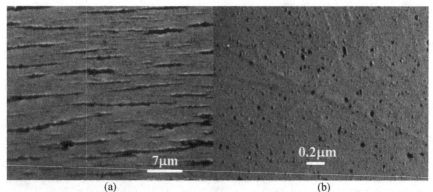

(a)   (b)

Figure 1. Optical pictures of the sample composites with (a) aligned and (b) randomly oriented CNT.

Sample characterization

The thermal characterization of the sample composites under considerations have been performed by means of the standard and reliable photothermal radiometry technique, whose experimental setup is described elsewhere[3]. A laser with a wavelength of 635 nm has been used to excited the samples and the photothermal radiometric signal has been recorded with a HgCdTe (Mercury-Cadmium-Telluride) detector (EG&G Judson J15D12-M204-S4), which is cooled with liquid nitrogen at the operation cryogenic temperature of 77 K. Before sending that signal to a digital lock-in amplifier (Stanford Research System Model SR850), this is amplified by a preamplifier with a frequency bandwidth 5 Hz-1 MHz (EG&G Judson Model PA-101), especially designed for operating with the HgCdTe detector. The thermal diffusivity α of the samples where then measured in the usual manner.

After determining the thermal diffusivity (α) of the sample and the specific heat capacity and density of the matrix and particles, the thermal conductivity (k) of the sample can be calculated by using the relation [10, 11]

$$k = \alpha\left(\rho_m c_m\left(1-f\right)+\rho_p c_p f\right), \tag{5}$$

where the subscripts m and p stands for the matrix and particles, respectively. The required specific heat and density of the matrix and CNT are summarized in Table 1.

Table 1: Properties of the matrix and CNTs.

Material	Specific heat $c\,(J/gK)$	Density $\rho(g/cm^3)$
Polyester resin (matrix)	1.06	1.16
CNT (particles)	0.64	2.3

DISCUSSION

The experimental data for the thermal conductivity of the composite with aligned and randomly oriented CNTs are shown in Fig.2(a), as a function of their volume fraction in comparison with the theoretical predictions of Eqs.(1)-(5). Note that for both cases, the predictions of the proposed approach are in good agreement with the experimental data, which exhibit a linear behavior. The thermal conductivity along the axis of the aligned CNTs is larger than the one obtained when they are randomly oriented, as expected. The thermal conductivity of the composite with aligned CNTs is about five times the thermal conductivity of the resin matrix at a particle concentration of just 4%. This is because the high thermal conductivity of the CNTs with respect to the one of the matrix. This enhancement due to the alignment decreases for large enough volume fractions of CNTs, as shown in Fig.2(b). This is reasonable, given that the interfacial effects between the CNTs and the matrix increase with the concentrations of particles.

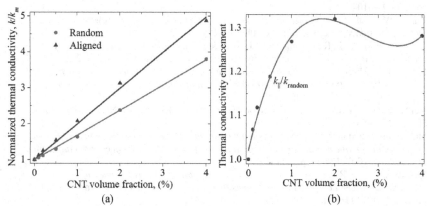

Figure 2. (a) Normalized thermal conductivity of the composites and (b) its enhancement. The Dots stand for experimental data and the continuous lines represent the theoretical predictions of Eqs.(1a) and (1b).

CONCLUSIONS

The thermal conductivity of composites made up of carbon nanotubes embedded in a polyester resin has been investigated based on an experimental and theoretical approach. The composite samples were prepared with a random and aligned distribution of carbon nanotubes. Its thermal conductivity was then measured by using the photothermal radiometry technique. The obtained experimental data has been accurately described by the proposed theoretical model, which takes into account the size effects of the nanotubes. It is expected that the obtained results can provide useful insights on the thermal design of composites based on carbon nanotubes.

ACKNOWLEDGMENTS

The technical assistance provided by M.S. Jose Bante Guerra and I.S.C. Georgina Espina Gurriz is highly acknowledged.

REFERENCES

1. Balandin, A.A., *Thermal properties of graphene and nanostructured carbon materials.* Nature Materials, 2011. **10**(8): p. 569-581.
2. Milton, G.W., *The Theory of Composites* 2002, Cambridge; New York: Cambridge University Press.
3. Zambrano-Arjona, M.A., R. Medina-Esquivel, and J.J. Alvarado-Gil, *Photothermal radiometry monitoring of light curing in resins.* Journal of Physics D: Applied Physics, 2007. **40**: p. 6098-6104.
4. Nan, C.W., et al., *Effective thermal conductivity of particulate composites with interfacial thermal resistance.* Journal of Applied Physics, 1997. **81**: p. 6692-6699.
5. Ordonez-Miranda, J., R.G. Yang, and J.J. Alvarado-Gil, *On the Thermal Conductivity of Particulate Nanocomposites* Applied Physics Letters, 2011. **98**(21): p. 233111-233113.
6. Lin, Z., L.V. Zhigilei, and V. Celli, *Electron-phonon coupling and electron heat capacity of metals under conditions of strong electron-phonon nonequilibrium.* Physical Review B, 2008. **77**(7): p. 075133-075149.
7. Korneva, G., et al., *Carbon nanotubes loaded with magnetic particles.* Nano Letters, 2005. **5**(5): p. 879-884.
8. Samouhos, S. and G. McKinley, *Carbon nanotube-magnetite composites, with applications to developing unique magnetorheological fluids.* Journal of Fluids Engineering-Transactions of the Asme, 2007. **129**(4): p. 429-437.
9. Medina-Esquivel, R.A., et al., *Thermal characterization of composites made up of magnetically aligned carbonyl iron particles in a polyester resin matrix.* Journal of Applied Physics, 2012. **111**(5): p. 054906-054913.
10. Almond, D.P. and P.M. Patel, *Photothermal Science and Techniques* 1996, London: Chapman & Hall.
11. Salazar, A., *On thermal diffusivity.* European Journal of Physics, 2003. **24**(4): p. 351-358.

Mater. Res. Soc. Symp. Proc. Vol. 1479 © 2012 Materials Research Society
DOI: 10.1557/opl.2012.1606

Experimental and theoretical studies of Boron Nitride Nanotubes: Electric arc discharge and DFT calculations.

R.A. Silva-Molina[1], R. Gámez-Corrales[2], and R.A. Guirado-López[3]

[1] Universidad Autónoma de San Luis Potosí, Doctorado Institucional de Ciencia e Ingeniería en Materiales, DICIM 78290, San Luis Potosí, S.L.P., México
[2] Departamento de Física, Universidad de Sonora, Rosales y Blvd. Luis Encinas 78000, Hermosillo, Sonora, México.
[3] Universidad Autónoma de San Luis Potosí, Instituto de Física, Zona Universitaria, 78290, San Luis Potosí, S.L.P., México
*Corresponding author: rogelio@correo.fisica.uson.mx
Telephone number: (52) 662-259-2114

ABSTRACT
We present a joint experimental and theoretical study dedicated to analyze the properties of Boron-Nitride (BN) nanotubes. First, multi-walled boron-nitride (MWBN) nanotubes were prepared by means of a modified electric arc discharge technique using boron-nitride powder. In a first stage, the BN powder was subjected to a ball milling process for about 100 hours in an atmosphere of ammonia. Later on, BN nanoparticle formation took place after the preparation of a pressed pellet at 300 °C to 25 kPa which was sintered in a furnace at approximately 1000 °C in nitrogen atmosphere for 15 hrs. The pellets were subsequently incorporated to the electrical arc discharge set up to obtain the MWBN nanotubes. The as-prepared MWBN nanotubes samples were characterized by scanning electron microscope, X-ray photoelectron spectroscopy, and micro RAMAN spectroscopy. Second, and in order to understand the measured data, extensive density functional theory calculations were performed. We present low energy atomic configurations for model finite-length armchair, zigzag, and chiral single-walled BN nanotubes, as well as for two-dimensional BN sheets. We calculate the vibrational spectra and the optical gap of each one of our considered structures and reveal how precise details of the local atomic environment can be revealed. Finally, we consider BN nanotubes functionalized with NH_2, glycine and S-H molecules. We present the structural characteristics of the adsorbed configurations, charge transfer effects, and the electronic behavior. We conclude by underlining the crucial role played by molecular functionalization in order to tune the properties of these kinds of systems.

Keywords: Nanotubes, RAMAN, Density Functional Theory.

INTRODUCTION

The hexagonal boron nitride is a semiconductor material with a ban gap 3.5-6 eV, those structures have been more studied for electronic structure by analogies of carbon nanotubes and optical properties. Recently, many studies shown the boron nitride tubes with a functionalized groups using a theoretical calculations like a Density Functional Theory DFT and shown how can changed the optical and electronics properties [1-2]. In this work we show different chirality for single-walled boron nitride nanotubes.

THEORY

Density functional theory (DFT), implemented by the Quantum Espresso program package is used to obtain the geometric and electronic structure of single-walled Boron Nitride (BNNT) nanotubes and functionalized BN nanotube and glycine. The double numerical basis sets plus polarization functional (DPN) were applied. All the calculations, in this work, were carried out using generalized-gradient approximation (GGA) [3-4]. Electron wave function was expanded in plane wave and the core treatment was used effective core potentials with a double numerical plus basis. Due to the charge transfer from boron to nitrogen, the tubular structure forms a dipolar shell and arranged molecular form inner boron of nanotube and the outer Nitrogen tube.

EXPERIMENTAL DETAILS

The nanotubes multiwall boron nitride is prepared by means of the experimental technique modified electrical discharge using boron nitride (BN) powder in hexagonal phase. In a first stage, the BN powder is subjected to a mechanical grinding process 100hrs in ammonia atmosphere. The preparation of nanoparticles of BN is performed after compacting BN pellets at 300°C and 25kPa and after that sintered in an oven at 1100°C in N_2 atmosphere for 15hrs. The pellets were then incorporated into the cell where electric shock was obtained multiwall BN nanotubes (BNNTs).

DISCUSSION

Figure 1 shows the optimized molecular structures, and because to the charge transfer from boron to nitrogen atom, the tubular structure forms a dipolar shell and arranged molecular form inner boron of nanotube and the outer nitrogen nanotube. Density isosurfaces of HOMO and LUMO of the 3 types of BN nanotubes chirality: a) Armchair, b) Zigzag and, c) Chiral single-walled Boron Nitride is depicted in figure 1. Armchair BNNT shows localized electrons, showing equiprobability density, contrary to the zigzag BN nanotubes which show overloaded molecules orbitals. Moreover, chiral single wall boron nitride nanotube shows a low probability density in the B, N atoms of the center of the nanotube, on the other hand the external B, N atoms of the NT show larger probability density [5-6].

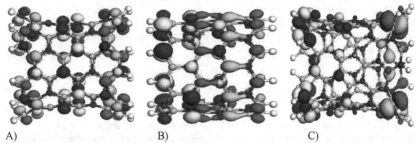

A)   B)   C)

Figure 1. - Optimized molecular structures and density isosurfaces of HOMO and LUMO in a) Armchair (6,6), b) Zigzag (6,0) and, c) Chiral (8,4) single-walled Boron Nitride nanotubes, respectively.

Figure 2 shows a single walled nanotube functionalized with two biomolecules of glycine optimized molecular structures and density isosurfaces of HOMO and LUMO [7]. Firstly, the molecule of glycine is placed near the extern wall of the nanotube; choosing different positions and configuration of the molecule with purpose to obtain the optimized structural condition in which the interaction is the most stable. Figure 2 is the most stable structure, and shows a localized deformation of the nearest atoms of the BNNT with the molecule of glycine, without affecting the eccentricity of the tube. At the same time, the electronic density shows localized electrons in the atoms of glycine which more probability than the atoms of the BNNT. The functionalizing with molecules of glycine reduces the band gap from 4.2 to 1.7eV.

Figure 2. - Optimized molecular structures and density isosurfaces of HOMO and LUMO of armchair single-walled Boron Nitride nanotubes glycine functionalized.

Figure 3 presents the Scanning Electron Microscopy (SEM) of BNNT obtained by means of the experimental technique modified electrical discharge using h-BN. The formation of a net of BN of few microns of length is clearly evidenced [8].

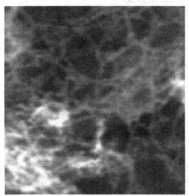

Figure 3. - SEM micro photography of BNNTs (100000x, 25kV) appreciating a nanotube network of several microns in length.

CONCLUSIONS

In this work we conduct experimental and theoretical study in order to analyze the structure and electronic properties of boron nitride nanotubes of type zigzag, armchair, and chiral. The nanotubes exhibit insulating properties, independent of diameter, chirality, or whether they are single wall or multi wall. The experimental results show a heterogeneous network with inclusions formed of boron nitride nanotubes, which were confirmed by Raman spectroscopy experimental results and theoretical calculations supported by density functional. These results give a good match between the theoretical and experimental vibrational normal modes. On the other hand, the BNNT glycine functionalized depicts a preference of the electronic density to the atoms of glycine, showing localized electrons. SEM microphotography shed light on the knowledge of the BNNT, evidencing the creation of a spider-like net of BNNT of several micrometers of length.

ACKNOWLEDGMENTS

The authors thank to the National Council of Science and Technology of Mexico (CONACyT) for its financial support, Universidad de Sonora. A scholarship from the CONACyT to R.A.S.M is gratefully acknowledged.

REFERENCES

1. A. Loiseau, *et al.*, "Boron Nitride Nanotubes with Reduced Numbers of Layers Synthesized by Arc Discharge," *Physical Review Letters*, vol. 76, pp. 4737-4740, 1996.
2. R. Arenal, "SYNTHÉSE DE NANOTUBES DE NITRURE DE BORE: ÉTUDES DE LA STRUCTURE ET DES PROPRI´ET´ES VIBRATIONNELLES ET ÉLECTRONIQUES," DOCTEUR DE L'UNIVERSIT´E ORSAY - PARIS SUD XI, Physique des Solides, Universit´e PARIS-SUD XI, Francia, 2005.
3. L. Wirtz, *et al.*, "Ab initio calculations of the lattice dynamics of boron nitride nanotubes," *Physical Review B*, vol. 68, p. 045425, 2003.
4. Y. F. Zhukovskii, *et al.*, "Ab initio simulations on the atomic and electronic structure of single-walled BN nanotubes and nanoarches," *Journal of Physics and Chemistry of Solids*, vol. 70, pp. 796-803, 2009.
5. G. Ciofani, *et al.*, "Preparation of Boron Nitride Nanotubes Aqueous Dispersions for Biological Applications," *Journal of Nanoscience and Nanotechnology*, vol. 8, pp. 6223-6231, 2008.
6. Johan Grasjo, *et al.*, "Local Electronic Structure of Functional Groups in Glycine as Anion Zwitterion and Cation in Aqueous Solution" *Journal Physical Chemistry*, 2009.
7. C. Zhi, *et al.*, "Immobilization of Proteins on Boron Nitride Nanotubes," *Journal of the American Chemical Society*, vol. 127, pp. 17144-17145, 2005.
8. P. Jaffrennou, *et al.*, "Optical properties of multiwall boron nitride nanotubes," *physica status solidi (b)*, vol. 244, pp. 4147-4151, 2007.

Mater. Res. Soc. Symp. Proc. Vol. 1479 © 2012 Materials Research Society
DOI: 10.1557/opl.2012.1607

Computational Fluid Dynamics in the Carbon Nanotubes Synthesis by Chemical Vapor Deposition

Alejandro Gómez Sánchez [1], Lada Domratcheva Lvova*[1], Víctor López Garza [1], Ramón Román Doval[1], María de Lourdes Mondragón Sánchez [2]
[1]Universidad Michoacana de San Nicolás de Hidalgo, Avenida Francisco J. Mujica S/N Ciudad Universitaria, C. P. 58030, Morelia, Michoacán, México.
[2]Instituto Tecnológico de Morelia, Avenida Tecnológico 1500, C. P. 58120, Morelia, Michoacán, México.

ABSTRACT

In this paper, an experimental study aimed at achieving better control of the deposition patterns of carbon nanotubes (CNTs) is presented. CNTs were grown on a long of reactor by the catalytic chemical vapor deposition (CVD) of a benzene/ferrocene solution at 1073 K. The deposition patterns on the substrate were controlled for process times and carrier gas flow rates. In order to investigate the reaction mechanism and production rate for the growth of CNTs in catalyst CVD, computational fluid dynamics (CFD) model was developed in this study. Then the computational model was integrated with the dynamic model to optimize the process parameters formulating a correlation between turbulence, deposition rate for the growth of carbon nanotubes and parameters as process time and carrier gas flow rate. Scanning electron microscopes (SEM) are used to characterize carbon nanotubes products.

Keywords: chemical vapor deposition (CVD), scanning electron microscopy (SEM), simulation.

INTRODUCTION

Today we need new materials to help us develop new technologies, nanotechnology helps us to manipulate the structures of materials and form others with very beneficial properties for the various applications of science. This study illustrates the application of CFD (computation fluid dynamic) by flow through a multi-step chemical reaction to predict the performance of carbon nanotubes in CVD (chemical vapor deposition) reactor. The objectives are simulated and analyze CFD in the synthesis of carbon nanotubes (CNTs) by CVD to optimize process parameters.

The CNTs deposition profiles inside a chemical vapor deposition reactor are strongly dependent on the reaction temperatures, process time, feed gas flow rates, carrier gas flow rates and reactor geometry. Obtaining and optimizations of the simulation parameters propose generate a high degree of purity in the samples. This type of optimization becomes more and more difficult with an increasing number of design parameters. The advent of computing power reaching at relatively low cost has spurred the development of numerical models.

One of the most important aspects in the production of carbon nanotubes is to control each and every one of the parameters and variables involved in the process. With the increase of parameters that need to be controlled and the complexity which entails, it is difficult to obtain an analytical solution to this problem and it is necessary to apply numerical methods. The use of these methods for parameter optimization in combination with the simulation is a promising method to overcome this problem. With these numerical and experimental results we can optimize the parameters of process of chemical vapor deposition of CNTs.

Carbon nanotubes have evolved into one of the most intensively studied materials and are responsible for co-triggering the Nanotechnology Revolution [1,2]. Applying the CFD calculation [3], total deposition rate of CNTs in the reactor is predicted with the variability of the processing parameters. A steady laminar flow coupled with heat transfer, was numerically solved for optimization of thermal CVD from a benzene/ferrocene solution in the reactor, it was compared with [4]. At first, through computational fluid dynamics, the rate of deposition was related with the process time and carrier gas flow rates.

EXPERIMENT

In this study, carbon nanotubes are produced by chemical vapor deposition (CVD) using two different process variables: time and flow from benzene and ferrocene. The growth of CNTs occurs within a reactor which is a horizontal pipe (1 inch of diameter) which passes through two furnaces. The first furnace is used to prevent flow turbulence caused by buoyancy effects and to sublime the ferrocene. In the second furnace CVD take place producing iron catalytic centers, in which CNTs grown. The molecules of benzene are organic precursors of carbon which are discomposed by pyrolysis at a temperature of 1073 K. Thermal decomposition of ferrocene provides Fe catalytic centers and a source of carbon. The reactor and equipment are represented in Figure 1.

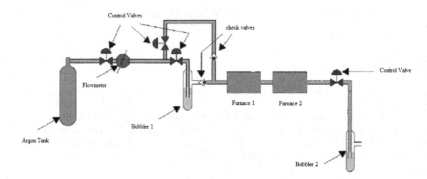

Figure 1. Diagram of the reactor and its components.

Argon flow is used as gas carrier and preventing combustion. Laminar flow conditions are recommendable to synthesize CVD products of better quality. The CNTs obtained were analyzed by scanning electron microscopy (SEM), which will display the size and structure of samples and their impurities.

It is simulated argon flow velocity profile and the temperature distribution. This will provide the conditions under which the process occurs and it can optimize the production of carbon nanotubes in the experimental tests. Conditions of simulation: Velocity profiles within the reactor with different variables; Reactions within the reactor; Solve settings and perform iterations; Check convergence; Examine the results.

The variables to be set for the experiments are shown in the Table I.

Table I. The variables to change were time and flow.

Time	1 and 1.5 hrs.
Temperature	1073 K
Flow (argon)	50, 75 and 100 ml/min
Ferrocene	500 mg
Benzene	200 ml

DISCUSSION

In this simulation were obtained velocity profiles, turbulence and concentrations of benzene. It is important in the synthesis conditions by CVD to have laminar flow along the second furnace where deposition of CNT's occurs in the tube walls. In this simulation we use the model k - ε for turbulence, because the geometry and the conditions above described. The distribution of velocities of flow in the reactor tube can be observed in the Fig. 2. The time needs to displace of oxygen is 5 minutes. By observing argon flow velocity in the second furnace it is possible to see the behavior of turbulence and then compare it with experimental results.

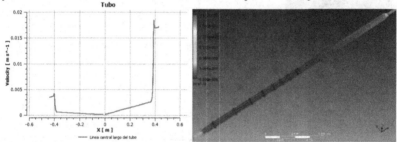

Figure 2. The distribution of velocity in the reactor tube.

The distribution of temperatures can be observed in the Figure 3. It can be observed that the temperature of the second furnace is increased turbulence and the velocity vectors, which can be seen in Figure 4 and Figure 5. The influence of turbulence in the second furnace has a relation with purity of CNTs. With lower turbulence there are better results with groups of aligned CNTs such as shown in Figure 5. It can be obtained better results. This means less impurity and larger nanotubes with lower velocity of flow as observed in Figure 5.

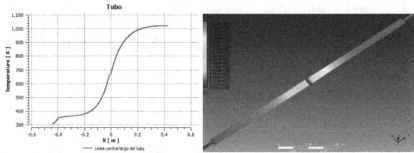

Figure 3. Distribution of temperatures in the reactor tube.

The Table II shows the experimental results of our six samples.

Table II. Amounts of carbon nanotubes obtained with variation of time and flow.

Time (hrs)	1			1.5		
Flow(ml/min)	50	75	110	50	75	110
CNTs (mg)	1716.6	570.5	977.5	1030.5	1486.2	1306.1

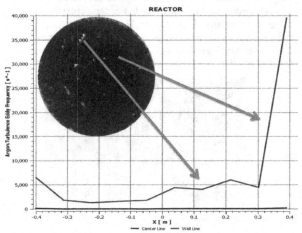

Figure 4. The turbulence around the wall where the nanotubes grow.

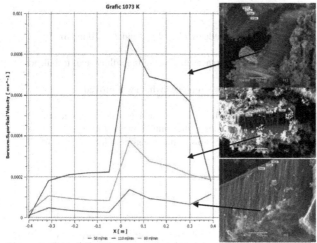

Figure 5. The relation between velocity and purity of the samples.

The results obtained by simulation were tested experimentally. It is demonstrated that the flows of 110 ml/min and major concentration of benzene it is produce more impurities and the minor lengths of CNT´s. Meanwhile the flows of 50 ml/min and lower concentration of benzene it is produce the lengths of CNT´s about of 400 μm and lower quantity of impurities, which can observe in the Figure 6.

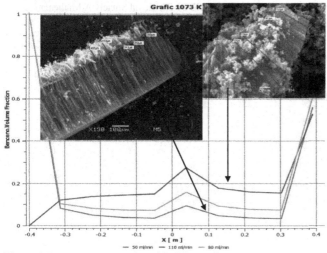

Figure 6. Benzene concentration with different flows.

CONCLUSIONS

- Multiwall CNTs were obtained with a longitude of 70-412 µm and 20-110 nm of diameter.
- It was simulated the flow through the two furnaces to observe the influence of turbulence and benzene volume fraction.
- Better quality of CNTs was obtained at lower turbulence which was proved experimentally.
- The best results of CNTs were with lower velocity of flow and as consequence lower benzene concentration, the fact which coincides in the simulation and the experiment.
- The model simulation allows find better conditions of the experiment for synthesis of CNTs of best quality.

ACKNOWLEDGMENTS

This work was supported by Coordination of Scientific Investigation of Michoacan University of San Nicolas de Hidalgo (UMSNH) and project Networks of Collaboration PROMEP (SEP), Mexico.

REFERENCES

1. C. S. Yah, G. S. Simate, K. Moothi, K. S. Maphutha and S. E. Iyuke, Trends in Applied Sciences Research, 6, 1270-1279 (2011)
2. M. Terrones, H. Terrones, Philosophical Transactions of The Royal Society A Mathematical Physical and Engineering Sciences, **361** (1813), 2789-806 (2004)
3. J. H. Fertziger and M. Peric, *Computational methods for fluid dynamics*, 2nd ed. (Springer, Berlin, 1999)
4. H. Endo, K. Kuwana, K. Saito, D. Qian, R. Andrews, E A. Grulke, Chem Phys Lett, 387, 307-11 (2004).

Contact author's email: ladadomracheva@yahoo.es

Mater. Res. Soc. Symp. Proc. Vol. 1479 © 2012 Materials Research Society
DOI: 10.1557/opl.2012.1608

Conductance Properties of Multilayered Silver-Mean and Period-Doubling Graphene Structures

G. Rodríguez-Arellano, D. P. Juárez-López, J. Madrigal-Melchor, J. C. Martínez-Orozco
and I. Rodríguez-Vargas
Unidad Académica de Física, Universidad Autónoma de Zacatecas, Calzada Solidaridad
Esquina con Paseo La Bufa S/N, 98060 Zacatecas, Zac., México.

ABSTRACT

In this work we alternate breaking-symmetry-substrates (BSS) and non-breaking-symmetry-substrates (NBSS) such as SiC and SiO_2, following the Silver-Mean (MSMGS) and Period-Doubling (MPDGS) sequences. We implement the Transfer Matrix technique to calculate the transmittance and the linear-regime conductance as a function of the most relevant parameters of the multilayered graphene structures: energy and angle of incidence, widths of BSS and NBSS regions and the generation of the quasi-regular sequence. We analyze the main difference of the transmission and conductance properties between MSMGS and MPDGS.

Keywords: layered structures, electronic structure, nanostructures

INTRODUCTION

Graphene has been the central topic in theoretical and experimental research, since year 2004 when it was experimentally isolated by Novoselov and Geim [1-2]. A central point in graphene is that electrons behave like relativistic particles [1-3], even when they move much slower that the speed of light, $v_F = c/300$. The consequences of this behavior are unusual effects such as minimum conductivity and Klein tunneling [4-5].

On the other hand, aperiodic order appears in different parts of nature, and it describes an increasing number of complex systems [6]. The most promising applications of quasi-regular multilayers have been as electronic and optic filters [7-8], specifically their most spectacular result in the field of nonlinear optics with the generation of second and third harmonics [8]. There are many studies on quasi-regular structures, particularly on Fibonacci, Cantor and Thue-Morse [9-13]. However, there are also others less investigated that could be of practical interest, like Silver-Mean and Period-Doubling [14-17].

Furthermore, the advent of new materials opens new possibilities both from the fundamental as technological standpoints. This is the case of graphene due to its outstanding properties and superb conditions for multiple technological applications [18-19]. From the fundamental point of view graphene serves as a natural bridge between condensed matter physics and high energy physics, offering the possibility to test relativistic effects on top-table experiments [5], as well as to open a new field in physics such as Relativistic Condensed Matter Physics, term that has being coined in the community. From the technological standpoint the opportunities are unprecedented taking into account that graphene gathers excellent thermal, electronic and mechanical properties,

Contact e-mail: jmadrigal.melchor@fisica.uaz.edu.mx

which make it one of the most serious candidates to replace silicon in electronics. Up to now, graphene has been subjected to extensive and intensive research practically in all relevant aspects of the material properties, see Ref. 18 and references there in.

Recently, we have studied the propagation properties of quasi-regular graphene structures [20]. The multilayered systems have been constructed alternating BSS and NBSS according to the Silver-Mean (MSMGS) and Period-Doubling (MPDGS) sequences, and then locating on them a graphene sheet. The BSS region generates a band gap [21], which results in potential barriers for electrons and holes in the multilayer structure. To complement our mentioned study, we calculate here the transport properties of Dirac electrons in Silver-Mean and Period-Doubling multilayered structures. Particularly, we carried out a comparative study of the linear-regime conductance between MSMGS and MPDGS.

THEORY

We consider a multilayered graphene structure of alternating q and k regions, where q represents the BSS region with wavevector $q = (q_x, q_y)$, with barrier width w_q. This region can be represented by the Hamiltonian $H=v_F\sigma\cdot p+t'\sigma_z$, with eigen-solutions,

$$\psi_q^\pm(x,y) = \frac{1}{\sqrt{2}}\binom{1}{v_\pm} e^{\pm iq_x x + iq_y y} \ , \qquad\qquad v_\pm = \frac{E-t'}{\pm q_x - iq_y} \ ,$$

where the plus and minus signs are the forward and the backward directions; here σ, v_F =$c/300$ and $t'=E_g/2$ are the Pauli matrices, Fermi velocity and the mass term ($t'= m\, v_F^{\,2}$) related to the energy gap.

In the case of NBSS region, with wavevector $k=(k_x, k_y)$ and width w_k, the Hamiltonian is $H=v_F\sigma\cdot p$, and its solutions,

$$\psi_k^\pm(x,y) = \frac{1}{\sqrt{2}}\binom{1}{u_\pm} e^{\pm ik_x x + ik_y y}, \qquad\qquad u(\vec{k},s) = se^{\pm i\theta},$$

where the plus and minus signs refers to waves propagating respectively, in the positive and negative directions of the x axis. The coefficients $u_\pm$ and $v_\pm$ are related directly to the angle of incidence [24]. More details about the notation and terminology can be found in [24].

Now, applying the boundary condition to the first order Dirac equation, i.e., the continuity of the wavefunction in interfaces of region q and k, we can obtain the relation for the corresponding coefficients

$$\begin{pmatrix} A_0 \\ B_0 \end{pmatrix} = D_0^{-1} \left(\prod_j (D_j P_j D_j^{-1}) \right) D_0 \begin{pmatrix} A_{N+1} \\ 0 \end{pmatrix},$$

where D_0 is the dynamical matrix of incident region and transmitted region, as well as D_j and P_j are the corresponding dynamical and propagation matrices of the j-th region of the multilayered structure, barriers and wells, corresponding to q and k respectively. This matrix are defined as

$$D_{j=k} = \begin{pmatrix} 1 & 1 \\ u_{+,j} & u_{-,j} \end{pmatrix}, \qquad P_{j=k} = \begin{pmatrix} e^{-ik_{x,j}w_j} & 0 \\ 0 & e^{ik_{x,j}w_j} \end{pmatrix},$$

$$D_{j=q} = \begin{pmatrix} 1 & 1 \\ v_{+,j} & v_{-,j} \end{pmatrix}, \qquad P_{j=q} = \begin{pmatrix} e^{-iq_{x,j}w_j} & 0 \\ 0 & e^{iq_{x,j}w_j} \end{pmatrix}.$$

Furthermore, defining the transfer matrix as [22]:

$$M = D_0^{-1} \left(\prod_j (D_j P_j D_j^{-1}) \right) D_0,$$

we can calculate the transmittance through the well known formula,

$$T = \frac{1}{|M_{11}|^2},$$

being M_{11} the first element of M. Using the transmittance or transmission probability we can obtain readily the linear-regime conductance through the Landauer-Buttiker formula

$$G/G_0 = E_F^* \int\limits_{-\pi/2}^{\pi/2} T(E_F^*, \theta) \cos\theta \, d\theta,$$

where $E_F^* = E_F/E_0$ is the dimensionless Fermi energy with $E_0 = t'$, $G_0 = 2e^2 L_y E_0/h^2 v_F$ is the fundamental conductance factor with L_y the width of the system in the transversal y-coordinate, and θ is the angle of the incident electrons with respect to the x-coordinate.

RESULTS AND DISCUSSION

To construct the Silver-Mean and Period-Doubling sequences we use the two-letter substitution rules $g(q)=qqk$ and $g(k)=q$ for Silver-Mean sequence and $g(q)=qk$ and $g(k)=qq$ for Period-Doubling sequence. We start at all cases with $g_1 =q$, first generation ($N=1$), that corresponds to a single barrier irrespective of the quasi-regular graphene structure used.

In Fig. 1 we show the transmittance as a function of the Dirac electron energy for MSMGS and MPDGS, top and bottom row respectively. We consider two generations of the sequences, N=2 and N=4, first and second columns, respectively. The parameters used are $w_q=30a$ and $w_k=50a$ and $t'=0.1$ eV at normal incidence, where a corresponds to the carbon-carbon distance of the hexagonal structure of graphene. As we can see, MSMGS have a richer transmittance-peak structure, even for the second generation. There are other interesting characteristics of the transmittance such as: 1) a richer structure as the generations increases, 2) multiple transmission peaks in the low energy range associated to Fabry-Pérot resonances and possibly bound states, 3) an envelope presented in practically all transmission spectra, specifically smooth for MPDGS, 4) the symmetry with respect to the origin of energies totally related to the equivalence between electrons and holes for massive particles in graphene, and 5) the different resonant peaks below and above the potential barrier energy, which can be bound states or scattering states [23].

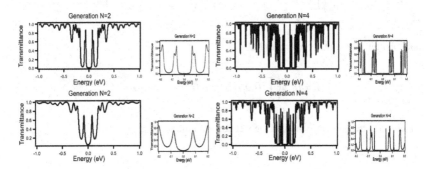

Fig. 1. Transmittance at normal incidence versus the electron energy for MSMGS (top row) and MPDGS (bottom row). The generations considered are, from left to right, 2 and 4 respectively. The widths of the gap and gapless regions used are $w_q=30a$ and $w_k=50a$ as well as the energy barrier height is $t'=0.1$ eV. The insets show the transmission probability at the low energy range.

Fig. 2 shows the transmittance as a function of energy and angle. The parameters considered are the same as in Fig. 1. MPDGS show more regions with total transmission than MSMGS, bright yellow region. We can also see zero transmission probability at normal incidence no matter the quasi-regular structure used as well as the generation considered. This transmission gap is related to the suppression of Klein tunneling as is the case of systems with BSS or massive Dirac electrons.

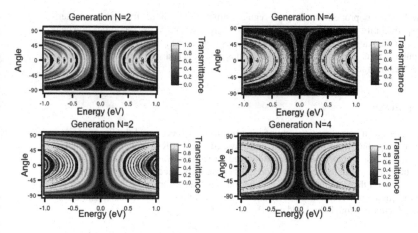

Fig. 2. (Color online). Transmittance as a function of energy and angle for (top row) MSMGS and (bottom row) MPDGS. The parameters used are $w_q=30a$ and $w_k=50a$ and $t'=0.1$ eV.

In Fig. 3 we show the conductance as a function of energy. We use the same values for w_q, w_k, and $t'=0.1$ eV. The most important difference between MSMGS and MPDGS is the richer peak structure sustained by MSMGS at the low energy range for $N=4$. These peaks can be related to resonant tunneling caused by confined states [24].

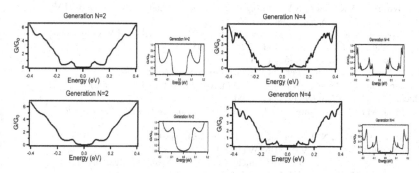

Fig. 3. Linear-regime conductance as a function of electron energy for MSMGS and MPDGS top and bottom row respectively. We present two generations $N=2$ and $N=4$ for each case. The parameters used for widths of the gap and gapless substrates are $w_q=30a$ and $w_k=50a$ and $t'=0.1$ eV.

Finally, it is important to point out that a deeper analysis of the transmission and conductance properties in quasi-regular graphene structures is needed in order to discern: 1) if the well known properties of quasi-regular structures such as fragmentation, self-similarity, fractality and critical wave function are presented in aperiodic graphene structures, 2) the nature of the conductance peaks as well as the role played on them by the quasi-regular effects, and 3) how to exploit the possible new and well known quasirregular characteristics in practical graphene devices.

CONCLUSIONS

In summary, a comparative study of the transmission spectra and the linear-regimen conductance for the Dirac electrons in multilayered Silver-Mean and Period-Doubling graphene structures is presented. The multilayered structure is obtained arranging SiC and SiO_2 substrates according to the substitution rules of the mentioned quasi-regular sequences and then sitting on them a graphene sheet. The transfer matrix formalism has been applied to the Dirac equations obtaining the well-known formulas for the transmittance and linear-regime conductance through the Landauer-Buttiker formula.

ACKNOWLEDGMENTS

The authors acknowledge the financial support given by the Consejo Nacional de Ciencia y Tecnología (CONACyT) through grant number CB-2010-151713.

REFERENCES

[1] K.S. Novoselov, A.K. Geim, S.V. Morozov, D. Jiang, Y. Zhang, S.V. Dubonos, I.V. Grigorieva and A.A. Firsov. *Science* **306**, 666-669 (2004).
[2] K.S. Novoselov, A.K. Geim, S.V. Morozov, D. Jiang, M.I. Katsnelson, I.V. Grigorieva, S.V. Dubonos and A.A. Firsov, *Nature* **438**, 197-200 (2005).
[3] P.R. Wallace. *Phys. Rev.* **71**, 622-634 (1947).
[4] Y.Zhang, Y.W. Tan, H.L. Stormer and P. Kim, *Nature* **438**, 201-204 (2005).
[5] M.I. Katsnelson, K.S. Novoselov and A.K. Geim, *Nat. Phys.* **2**, 620-625 (2006).
[6] E. Macia, *Rep. Prog. Phys.* **69**, 397-441 (2006).
[7] V. Agarwal, M.E. Mora-Ramos and B. Alvarado-Tenorio, *Phot. Nano. Fund. Appl.* **7**, 63-68 (2009).
[8] S.N. Zhu, Y.Y. Zhu and N.B. Ming, *Science* **278**, 843-846 (1997).
[9] J. M. Luck. *Phys. Rev. B.* **39**, 9, 5834-5849 (1989).
[10] R. Rodríguez-González, J.C. Martínez-Orozco, J. Madrigal-Melchor and I. Rodríguez-Vargas, *AIP Conf. Proc. LDSD*, (2011). To be published.
[11] F.F. de Madeires, E.L. Albuquerque and M.S. Vasconcelos, *J. Phys.: Condens. Matter.* **18**, 8737–8747 (2006).
[12] W. Steurer and D.S. Widmer. *J. Phys. D: Appl. Phys.* **40**, R229–R247 (2007).
[13] A. Ghosh and S.N. Kamakar. *Phys. Rev. B.* **58**, 2586–2590 (1998).
[14] I. Gahramanov and E. Asgerov. *Arxiv: 1010.2476v3 [cond-mat.-nn]*, (2012).
[15] V.Z. Cerovski, M. Schreiber and U. Grimm, *Phys. Rev. B.* **72**, 54203 (2005)
[16] H. Aynaou1, E.H. El Boudouti1, Y. El Hassouani1, A. Akjouj2, B. Djafari-Rouhani, J.

Vasseur, A. Benomar, and V. R. Velasco. *Phys. Rev. E.* **72**, 056601 (2005)

[17] H. Rahimi and S.R. Entezar, *Physica B.Condens. Matter.* **406**, 17, 3322-3327 (2011).

[18] A.H. Castro-Neto, F. Guinea, N.M.R. Peres, K.S. Novoselov and A.K. Geim, *Rev. Mod. Phys.* **81**, 109-162 (2009).

[19] A.K. Geim, and K.S. Novoselov, *Nature Mater.* **6**, 183-191 (2007).

[20] G. Rodríguez-Arellano, D.P. Juárez-López, J. Madrigal-Melchor, R. Pérez-Álvarez, J.C. Martínez-Orozco and I. Rodríguez-Vargas, (*Mater. Res. Soc. Symp. Proc.* **1371**, Cancun, Mexico, 2012).

[21] S.Y. Zhou, G.H. Gweon, A.V. Fedorov, P.N. First, W.A. de Heer, D.H. Lee, F. Guinea, A.H. Castro-Neto and A. Lanzara, *Nat. Mater.* **6**, 770-775 (2007).

[22] P. Yeh, *Optical waves in layered media*, (John Wiley & Sons, Inc., New Jersey, 2005).

[23] S. Datta, *Electronic Transport in Mesoscopic Systems*, (Cambridge University Press, 1995).

[24] J. Viana Gomes and N.M.R. Peres, *J. Phys.: Condens. Matter*, **20**, 325221 (2008).

Mater. Res. Soc. Symp. Proc. Vol. 1479 © 2012 Materials Research Society
DOI: 10.1557/opl.2012.1609

Nonlinear absorption coefficient and relative refraction index change for an asymmetrical double delta-doped quantum well in GaAs with a Schottky barrier potential.

J. G. Rojas-Briseño[1], J. C. Martínez-Orozco[1], I. Rodríguez-Vargas[1], C. A Duque[2] and M. E. Mora-Ramos[2,3].

[1]Unidad Académica de Física. Universidad Autónoma de Zacatecas. Calzada Solidaridad esquina con Paseo la Bufa S/N, C.P. 98060. Zacatecas, Zac. México.
[2]Instituto de Física, Universidad de Antioquia, AA 1226, Medellín, Colombia.
[3]Facultad de Ciencias, Universidad Autónoma del Estado de Morelos, Ave. Universidad 1001, C.P. 62209, Cuernavaca, Morelos, México.

ABSTRACT

Semiconductor devices have been improved by using delta-doped quantum well (DDQW) of impurities due to the great amount of charge carriers it provides. The first proposals consisted of a DDQW close to the Schottky barrier potential in the gate terminal in a FET [1]. In this work we reported the energy levels spectrum for n-type double-DDQW with a Schottky barrier (SB) at their neighborhood in a Gallium Arsenide (GaAs) matrix. In addition to consider only the linear optical approximation we take into account the third order correction to the absorption coefficient and the refractive index change. We report those properties as a function of the Schottky Barrier Height (SBH), several separation distances between the DDQWs, and hydrostatic pressure effects. The results shown that the magnitude of intensity resonance peaks are controlled by the asymmetry of the DDQW+SB.

Keywords: electronic structure, optical properties, III-V.

INTRODUCTION

Spatial localization of impurities in a single monolayer of a semiconductor crystal represents the ultimate physical limit of dopant distributions. Delta-doping is one of several terms used to name it [1-5]. Schubert and Ploog [6] proposed the δ-FET, a system where the conduction channel between the drain and the source is formed by a δ-doped layer. It is a system consisting of a metal-semiconductor and a DDQW, with the quasi-two-dimensional electronic cloud of the well being affected by the electric field of intensity V_c/d, where V_c is the height of the Schottky barrier and d is the distance separating the δ-well from the barrier. Self-consistent as well as analytical modeling of delta-doped field effect transistors were put forward by Gaggero-Sager et al. [7,8]. In the latter work, the analytical description of the confining potential profile was given along the lines of the Thomas-Fermi approximation introduced by Ioriatti [9]. The nonlinear optical properties in this kind of structures had not been investigated until very recently. Martínez-Orozco et al. studied the optical absorption and the relative change in the refractive index in a n-type δ-doped GaAs FET [10]. The aim of the present work is to extend such an investigation to the case of a GaAs n-type δ-FET-like configuration, with an asymmetric double delta well configuration, looking for a possible amplification of the nonlinear optical response in these systems, and taking into account the effect of applied hydrostatic pressure.

THEORY

Here we are considering the problem of an electron confined within an asymmetric n-type double-DDQW in a GaAs matrix with a close Metal-Semiconductor contact characterized though a Schottky barrier potential. The first n-type DDQW is located at $z = 0$ the second one is at $z = d$. This means that the potential energy function is taken as a combination the parabolic-type Schottky barrier contribution together with the double V-shaped well geometry typical for the delta-doped systems. This part of V (z) is analytically written within the one-dimensional local-density Thomas-Fermi approximation [8,9]. Our model of potential for the conduction band is given by:

$$V(z) = \frac{4\pi}{\varepsilon_r} N_d (z - l)^2 - \frac{\alpha^2}{(\alpha|z| + z_1)^4} - \frac{\alpha^2}{(\alpha|z - l_p| + z_2)^4} \tag{1}$$

this expression is given in atomic effective units. N_d represents the background impurities density, which has been taken with a value of $N_d = 1.0 \times 10^{18}$ cm^{-3}. ε_r represents the static dielectric constant for the host material, that in general depends on the applied hydrostatic pressure (P). In this expression $\alpha = 2/15\ \pi$ and z_i $(i=1,2)$ represents a fundamental length scale for the problem that depends on the value of the two dimensional density of impurities associated to each DDQW as follows:

$$z_i = \left(\frac{\alpha^3}{\pi N_{2d}^i} \right) \tag{2}$$

As we already mentioned above, in this work we take into account the hydrostatic pressure effect on the system through the dependency of the electron effective mass $m(P)$ and dielectric permittivity constant $\varepsilon_r(P)$ [10], that are given by:

$$\frac{m(P)}{m_0} = \left[1 + \frac{15020}{1519 + 10.7P} + \frac{7510}{1519 + 10.7P + 341} \right]^{-1} \tag{3}$$

$$\varepsilon_r(P) = 12.82\ e^{(-1.67\times 10^{-3}\ kbar^{-1}P)} \tag{4}$$

In this case we also need to include the effect that the hydrostatic pressure has onto the Schottky barrier height (SBH) through the following functional dependency [11]

$$V_c = V_c(0) + \beta P + \gamma P^2 + \kappa P^3, \tag{5}$$

where the parameters have the values $\beta = 11.21$ meV/kbar, $\kappa = -0.345$ meV/kbar^2, and $\gamma = 0.25$ meV/kbar^3. It must be stressed that we limit our work to the low pressure regime, within the range of 0 to 6 kbar, because this experimental expression (5) is reported to be valid only within this hydrostatic pressure range.

We obtain the energy levels and the corresponding wave functions for single-electron states with the use of the method developed by Xia and Fan [12]. Then, we are able to compute the intersubband optical properties for the system with the help of the expressions reported by Yildirim [13] for the linear and nonlinear intersubband optical absorption coefficient:

$$\alpha^{(1)}(\omega) = \omega e^2 \sqrt{\frac{\varepsilon}{\mu}} \left[\frac{\rho \hbar \Gamma_{10} |M_{10}|^2}{(E_{10} - \hbar\omega)^2 + (\hbar\Gamma_{10})^2} \right], \qquad (6)$$

$$\alpha^{(3)}(\omega, I) = -\omega e^4 \sqrt{\frac{\varepsilon}{\mu}} \left(\frac{1}{2n\varepsilon_0 c} \right) \left[\frac{\rho \hbar \Gamma_{10} |M_{10}|^2}{[(E_{10} - \hbar\omega)^2 + (\hbar\Gamma_{10})^2]^2} \right]$$
$$\left\{ 4|M_{10}|^2 - \frac{|M_{11} - M_{00}|^2 [3E_{10}^2 - 4E_{10}\hbar\omega + \hbar^2(\omega^2 - \Gamma_{10}^2)]}{E_{10}^2 + (\hbar\Gamma_{10})^2} \right\}, \qquad (7)$$

the total absorption coefficient is given by

$$\alpha(\omega, I) = \alpha^{(1)}(\omega) + \alpha^{(3)}(\omega, I). \qquad (8)$$

with respect to the relative refractive index, the linear contribution is given by

$$\frac{\Delta n^{(1)}(\omega)}{n} = \frac{\rho e^2 |M_{10}|^2}{2n^2 \varepsilon_0} \left[\frac{E_{10} - \hbar\omega}{(E_{10} - \hbar\omega)^2 + (\Gamma_{10}\hbar)^2} \right], \qquad (9)$$

whilst the corresponding third-order nonlinear contribution is

$$\frac{\Delta n^{(3)}(\omega, I)}{n} = -\frac{\rho e^4 |M_{10}|^2}{4n^3 \varepsilon_0} \frac{\mu c I}{[(E_{10} - \hbar\omega)^2 + (\Gamma_{10}\hbar)^2]^2}$$
$$\left[4(E_{10} - \hbar\omega)|M_{10}|^2 - \frac{(M_{11} - M_{00})^2}{(E_{10})^2 + (\Gamma_{10}\hbar)^2} \{(E_{10} - \hbar\omega)[E_{10}(E_{10} - \hbar\omega) - (\Gamma_{10}\hbar)^2] - (\Gamma_{10}\hbar)^2(2E_{10} - \hbar\omega)\} \right] \qquad (10)$$

Finally the total relative refractive index change is given by

$$\frac{\Delta n(\omega, I)}{n} = \frac{\Delta n^{(1)}(\omega)}{n} + \frac{\Delta n^{(3)}(\omega, I)}{n}. \qquad (11)$$

In these expressions we have that the matrix elements appearing are given by $M_{if} = <\psi_i|z|\psi_f>$, whereas $E_{10} = E_1 - E_0$, is the main intersubband optical transition. $\Gamma_{10}\hbar$ is the damping term associated with the lifetime of the electrons due to intersubband scattering, I is the intensity of the optical radiation in the system, and the other are well known physical parameters.

RESULTS AND DISCUSSION

First of all we calculate the electronic level structure for the system as a function of the relative distance between DDQWs, the Schottky barrier height and, finally we study the dependence with the hydrostatic pressure. To generate the results shown in the Fig. 1, a fixed value of ionized impurity density in the first DDQW, $N_{2d}^1 = 7.5 \times 10^{12}$ cm^{-2}, is taken and the value for the second DDQW is set as $N_{2d}^2 = 5.0 \times 10^{12}$ cm^{-2}. The SBH is 300 meV. In the plot we report four energy levels with their corresponding wave functions. It is important to note that the ground state is located in the first DDQW meanwhile the first exited state is at the second, at least for this particular set of parameters. We obtained from our calculations that if the second DDQW has a smaller value of impurities per cm^{-2} the electronic structure is almost completely confined within the first DDQW. If N_{2d}^2 is of the order of N_{2d}^1, each well can bound two energy levels. Such a bound states distribution has implications on the optical properties reported in this work.

Figure 1: Potential profile and electronic level structure for the double-DDQW+SBH system. In this figure we observe that we have at least four energy levels and that the first excited state is a bound state of the second DDQW. The set of parameters are: $N_{2d}^1 = 7.5 \times 10^{12}$ cm^{-2}, $N_{2d}^2 = 5.0 \times 10^{12}$ cm^{-2}, The contact potential is $V_c = 300$ meV, and the relative distance between wells in of $d = 20$ nm.

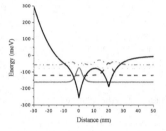

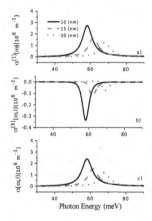

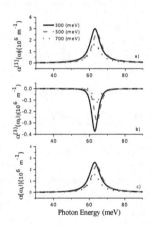

Figure 2: Absorption coefficient for the double-DDQW+SBH. In a) we plot the linear absorption coefficient for $d = 10$ nm (Continuous line), $d = 15$ nm (dashed-line) and $d = 20$ nm (doted-line). In b) and c) we report the third order correction and the total absorption coefficient, respectively.

Figure 3: Absorption coefficient for the double-DDQW+SBH. In a) we plot the linear absorption coefficient for $V_c = 300$ meV (Continuous line), $V_c = 500$ meV (dashed-line) and $V_c = 700$ meV (doted-line). In b) and c) we report the third order correction and the total absorption coefficient, respectively.

With regard to the absorption coefficient, we first investigate the absorption peak behavior as a function of the separation between wells by fixing the other system parameters. In this particular case the value used for the first-DDQW impurity density is $N_{2d}^1 = 6.0 \times 10^{12}$ cm^{-2}. The second-DDQW impurities density in this case is of $N_{2d}^2 = 2.0 \times 10^{12}$ cm^{-2}. In the figure 2 a) we plot the linear absorption coefficient as a function of the incident light frequency for relative inter-well distances of 10, 15 and 20 nm. We found that as the relative distance increases, the absorption coefficient peak exhibits a blueshift as well as a decrement in the intensity peak. This is fundamentally due to the fact that with the growth in the separating distance, the first excited level tends to locate in the first potential well (keep in mind that N_{2d}^2 is small enough). Also, overall effective confining well width decreases and the basic intersubband transition energy difference augments. However, the magnitude of the associated dipole matrix element is significantly reduced given that there will be an increase in the wave function's localization around the origin. In figure 2 b) we report the calculated third order correction for an intensity of $I = 1.0$ MW/cm^2 and this correction is about 10 % of the linear absorption coefficient. Finally in the figure 2 c) report the total absorption coefficient for the system. The same physical arguments explain the corresponding variations depicted.

We also investigate the effect that the contact potential will have on these properties. The results are presented in the figure 3, for three values of V_c. Basically, we found that the absorption peak energy position remains unchanged but that the intensity decrease as we raise the Schottky barrier potential, as a result of the reduction in the magnitude of the transition dipole matrix element.

The calculated relative refractive index change for the system is shown as a function of the relative distance between wells in the graphics of the left column of figure 4, as well as a function of the contact potential, as reported on the right column of figure 4. The system parameters are the same as those used in figure 2 and 3. The corresponding changes of the resonant peak position and amplitude are justified by the discussion made regarding the results appearing in the figure 2.

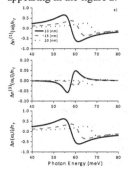 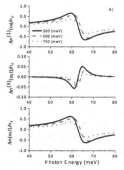

Figure 4: Relative refractive index change as a function of a) the relative separation distance between wells, and b) as a function of the contact potential barrier height. We report the linear (top-plot), third-order correction (middle-plot), and the total relative refractive index change (bottom-plot), respectively. In general is more important the relative distance between wells than the contact potential.

Finally, in the figure 5, we present the absorption coefficient and relative refractive index change as functions of the applied hydrostatic pressure (P) in the low pressure regime. This calculation was performed for the same set of parameters of the impurities density of the DDQWs as in the previous figures but the separation distance in this case is set as $d = 15$ nm, and the contact potential is fixed to $V_c = 300$ meV. In general, it is found that the hydrostatic pressure slightly increases the linear and nonlinear absorption coefficient's amplitude with the consequent rise in the corresponding total absorption coefficient. This slight increase has to do with the increment in the transition dipole matrix element due to the pushing effect of the increasing Schottky barrier height on the electron probability density. With respect to the relative refractive index change no significant modifications due to applied hydrostatic pressure are found as well.

Figure 5: a) Absorption coefficient and b) relative refractive index change as a function of photon energy for different values of the applied hydrostatic pressure: $P = 0$ kbar (Solid-line), $P = 2$ kbar (Dashed-line) and $P = 4$ kbar (Dotted line). The parameters in these calculations are $d = 15$ nm and the contact potential fixed to $V_c = 300$ meV.

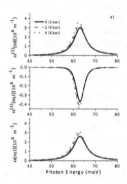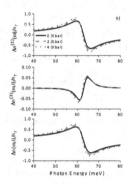

CONCLUSIONS

In this work we report the theoretical determination of the main intersubband (E_{10}) optical absorption coefficient and the corresponding relative refractive index changes for a GaAs double-DDQW+SBH system as a function of several separation distances between wells (d), contact potential (V_c) and hydrostatic pressure (P). These calculations were done within the effective mass approximation with an analytical self-consistent expression for the n-type DDQW based in the Thomas-Fermi approximation and using also the depletion approximation for the metal-semiconductor contact. In general, it is found that as long as we diminish the relative separation of the second DDQW, the absorption maximum peak experiences a blueshift as well as an increase in the absorption coefficient intensity. Also, when one increases the contact potential value, the absorption coefficient decreases, but the maximum of the resonant peak keeps, practically, the same energy position. We found also that the hydrostatic pressure effects can increase the amplitude for the absorption coefficient. An analogous behavior can be observed in the case of the relative change in the refractive index.

ACKNOWLEDGMENTS.

The authors would like to acknowledge to the National Science and Technology Council (CONACyT) from Mexico and to the Administrative Department of Science, Technology and Innovation (COLCIENCIAS) from Colombia for the partial financial support trough bilateral collaboration grant B330.500. MEMR is grateful to CONACYT for support under sabbatical Grant 2011-2012 No. 180636.

REFERENCES

1. G. H. Döhler, Surf. Sci. **73**, 97 (1978).

2. C. E. Wood, G. Metze, J. Berry and L. F. Eastman, J. Appl. Phys. **51**, 583 (1980).

3. M. Zachau, F. Koch, K. Ploog, P. Roentegen and H. Beneking, Sol. Stat. Commun. **59**, 591 (1986).

4. M. E. Lazzouni and L. J. Sham, Phys. Rev B **48**, 8948 (1986).

5. R. Van de Walle, R. L. Van Meirhaeghe, W. H. Laflere and F. Cordon, J. Appl. Phys. **74**, 1885 (1993).

6. E. F. Schubert and K. Ploog. Jpn. J. Appl. Phys. **24**(8), L608 (1985).

7. L. M. Gaggero-Sager and R. Pérez-Alvarez, J. Appl. Phys. **78**, 4566 (1995).

8. L. M. Gaggero-Sager and M. E. Mora-Ramos, Mat. Sci. Eng. B **47**, 279 (1997).

9. L. Ioriati, Phys. Rev. B **41**, 8340 (1990).

10. J.C. Martínez-Orozco, M.E. Mora-Ramos and C.A. Duque, Phys. Status Solidi B, **249**, 146 (2012)

11. G. Çankaya, N. Uçar, E. Ayyildiz, H. Efeoğlu, A. Türüt, S. Tüzemen, and Y. K. Yoğurtçu. Phys. Rev. B. **60**, 15944 (1999).

12. J.-B. Xia and W.-J. Fan, Phys. Rev. B. **40**, 8508 (1989).

13. H. Yildirim and M. Tomak, Eur. Phys. J. B **50**, 559564 (2006).

Mater. Res. Soc. Symp. Proc. Vol. 1479 © 2012 Materials Research Society
DOI: 10.1557/opl.2012.1610

Nonlinear optical properties related to intersubband transitions in asymmetrical double δ-doped GaAs; effects of an applied electric field

K. A. Rodríguez-Magdaleno[1], J. C. Martínez-Orozco[1], I. Rodríguez-Vargas[1], M. E. Mora-Ramos[2,3], C.A. Duque[3].

[1] Unidad Académica de Física. Universidad Autónoma de Zacatecas. Calzada Solidaridad esquina con Paseo a La Bufa S/N. C.P. 98060, Zacatecas, Zacatecas. México.
[2] Facultad de Ciencias. Universidad Autónoma del Estado de Morelos. Ave. Universidad 1001, C.P. 62209, Cuernavaca, Morelos, México.
[3] Instituto de Física, Universidad de Antioquia, AA 1226, Medellín, Colombia.

ABSTRACT

In this work, we calculated the ground and first excited states of an electron confined in an asymmetric double DDQW system within a Gallium Arsenide (GaAs) matrix. The two-dimensional impurities density (N_{2d}) considered in our calculation are within the range of 10^{12} to 10^{13} cm^{-2}. We obtain the linear and nonlinear optical properties related to intersubband transitions as a function of the spacing between δ-doped wells, two-dimensional impurities concentrations as well as in presence of electric field. We reported results for the linear and nonlinear optical absorption coefficient and in the relative refractive index changes. Our results show that the asymmetry induced in the double δ-doped well system gives rise to values that are several orders of magnitude higher in the resonant peaks intensity.

Keywords: electronic structure, optical properties, III-V.
Corresponding author: karelyrod@fisica.uaz.edu.mx

INTRODUCTION

Between the quantum confined systems that have attracted much attention since its proposal in 1980 by Wood et el. [1] are the δ-doped quantum wells (DDQW) of impurities. This consists in the incorporation of a huge amount of impurities, donors (n-type) or acceptors (p-type), in a single layer of semiconductor material. A self-consistent potential model widely used to study such systems was proposed by Ioriatti in 1990 [2]. The potential energy function that corresponds to the confining potential profile in a double DDQW is another asymmetric configuration that can provide non-negligible contributions from the diagonal expectation values of the dipole moment operator. Up to these author's knowledge, this configuration has not been previously considered in the study of optical effects such as the nonlinear optical absorption (NOA) and refractive index corrections, although there is a previous report on optical nonlinearities in δ-doped AlGaN/GaN quantum well heterostructures [3], as well as two very recent ones dealing with nonlinear optical responses in δ-doped field effect transistors [4, 5]. Thus, the aim of this study is to present some initial results on the linear and nonlinear optical absorption and relative change of the refractive index in GaAs-based asymmetric double DDQW with and without the application of a stationary electric field.

THEORY

Our potential model for the conduction band profile based in a Thomas-Fermi self consistent potential obtained by Ioratti [2] and is given by:

$$V_n = \frac{\alpha^2}{(\alpha|z - l_p| + z_0)^4} + \frac{\alpha^2}{(\alpha|z + l_p| + z_0)^4}.$$

The expressions α and z_0 are given in [2, 4]. In particular, the quantity z_0 depends on the two-dimensional concentration of ionized impurities (N_{d1} and N_{d2}, in each case), which precisely serves as the origin of the asymmetry as can be seen in figure 1 below. We have chosen to consider a fixed value of the ionized impurity density in the left-hand DDQW while the value of N_{d2} is varied. Once we obtain the electronic structure by solving the conduction band effective mass Schrödinger-like equation we are able to calculate the total absorption coefficient and the relative refractive index changes. First, the total absorption coefficient is given by [4, 6],

$$\alpha(\omega, I) = \alpha^{(1)}(\omega) + \alpha^{(3)}(\omega, I),$$

where, $\alpha^{(1)}(\omega)$ is the first order contribution

$$\alpha^{(1)}(\omega) = \omega e^2 \sqrt{\frac{\varepsilon}{\mu}} \left[\frac{\rho \hbar \Gamma_{10} |M_{10}|^2}{(E_{10} - \hbar\omega)^2 + (\hbar\Gamma_{10})^2} \right],$$

and $\alpha^{(3)}(\omega, I)$ is the third nonlinear correction,

$$\alpha^{(3)}(\omega, I) = -\omega e^4 \sqrt{\frac{\varepsilon}{\mu}} \left(\frac{I}{2n\epsilon_0 c} \right) \frac{\rho \hbar \Gamma_{10} |M_{10}|^2}{(E_{10} - \hbar\omega)^2 + (\hbar\Gamma_{10})^2} \left\{ 4|M_{10}|^2 \right.$$
$$\left. - \frac{|M_{11} - M_{00}|^2 |3E^2_{10} - 4E_{10}\hbar\omega + \hbar^2(\omega^2 - \Gamma^2_{10})|}{(E_{10})^2 + (\hbar\Gamma_{10})^2} \right\}.$$

With respect to the relative refractive index changes, we have

$$\frac{\Delta n(\omega, I)}{n} = \frac{\Delta n^{(1)}(\omega)}{n} + \frac{\Delta n^{(3)}(\omega, I)}{n},$$

where

$$\frac{\Delta n^{(1)}(\omega, I)}{n} = \frac{\rho e^2 |M_{10}|^2}{2n^2\epsilon_0} \left[\frac{E_{10} - \hbar\omega}{(E_{10} - \hbar\omega)^2 + (\hbar\Gamma_{10})^2} \right]$$

134

and the correspondingly third order correction is given by

$$
\frac{\Delta n^{(3)}(\omega)}{n} = -\frac{\rho e^4 |M_{10}|^2}{4n^3 \epsilon_0} \frac{\mu c I}{|(E_{10} - \hbar\omega)^2 + (\hbar\Gamma_{10})^2|^2} \left[4(E_{10} - \hbar \right.
$$
$$
\left. -\frac{(M_{11} - M_{00})^2}{(E_{10})^2 + (\hbar\Gamma_{10})^2} \{(E_{10} - \hbar\omega) |(E_{10} - \hbar\omega) - (\hbar\Gamma_{10})^2| - (\hbar\Gamma_{10})^2(2E_{10} \right.
$$
$$
\left. - \hbar\omega)\} \right].
$$

In those expressions, $M_{ij} = \langle \phi_i | z | \phi_j \rangle$, where ϕ_i and ϕ_j correspond to the wave functions of the first two states that are obtained by solving the Schrodinger equation. $E_{10} = E_1 - E_0$ is the energy difference of the intersubband transition, $\hbar\Gamma_{10} = \hbar/T_{10}$ is the damping term associated with the lifetime of electrons due to the dispersion intersubband I (= 0.1MW/cm^2) is the intensity of optical radiation in the system, which relates the electric field of the incident light at through the expression

$$
I = \frac{2n}{\mu c} |E(\omega)|^2.
$$

The other parameters are well known physics quantities.

RESULTS AND DISCUSSION
   First we obtain the electronic structure for the system as function of the spacing between δ-doped wells, the two-dimensional impurities density as well as in the presence of electric field. In all cases we fix the impurities density for the first well at N_{2d1}=7.5, from now and on all the two dimensional impurities densities are in units of 10^{12} cm^{-2}. Using the calculated energy states we obtain the absorption coefficient and the relative refractive index changes taking into account the third order correction in both cases. We report these properties for different values of the impurities density of the second δ-doped well, as well as a function of the applied electric field.

   Figure 1 contains a schematic picture of the conduction band energy profile of the double DDQW together with the corresponding confined single-electron eigenstates in the zero applied field case [Fig. 1 a)]. Also, the linear, third-order nonlinear and total optical absorption coefficient [Fig. 1 b)] and the linear, third-order nonlinear and total relative change of the refractive index [Fig. 1 c)] are depicted for the cases of N_{d2}=2.5; 2.8. It can be seen that the slight increment in N_{d2} from 2.5 to 2.8 has a significant repercussion on the amplitude of the resonant peaks in both optical responses, together with a shift towards the lower frequencies of the incident light (redshift). The figure 1 a) helps us to understand this latter fact by observing that the increase of N_{d2} implies a deepening of the right-hand QW, which has the consequence of the reduction between the two energy levels involved in the E_{10} transition.

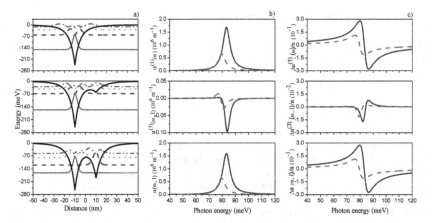

Figure 1. Potential energy profile and electronic structure for the system a), linear and nonlinear optical responses b) and c) in a GaAs-based double DDQW. Here we show the probability density for $N_{d2} = (0.1, 1.0,$ and $5.5)$. b) Absorption Coefficient, $\alpha^{(1)}$ (top), $\alpha^{(3)}$ (middle), and total, (bottom), c) Relative change of the refractive index. $\Delta n^{(1)}/n$ (top), $\Delta n^{(3)}/n$ (middle) and total (bottom).

The decrease in the amplitudes of these optical coefficients when there is an increment on N_{d2} is related with the fall in the value of the absolute dipole matrix element of the transition $|M_{10}|^2$. As long as the potential well configuration more closely resembles that of a symmetric double DDQW, the value of this matrix element will reduce due to confined state symmetry reasons. It is possible to observe an apparent "anomalous" behavior of the nonlinear corrections to the optical absorption which reflects in positive values of this quantity along a certain interval of the incident photon energy. One readily observes that the corresponding mathematical expression have a minus factor which, under normal conditions, will make the whole third-order contributions to be negative. In our case, it seems that there is a range of frequencies for which the term between braces in $\alpha^{(3)}$ become negative and render an overall positive value for this nonlinear coefficient. A further study is needed in order to reveal the reason of this rather surprising result.

Figure 2 shows the effects of the application of a stationary electric field on the confining potential profile [Fig 2 a)], the optical absorption coefficient [Fig. 2 b)], and the relative change of the refractive index [Fig. 2 c)], in a GaAs asymmetric double DDQW. In this case, the value of the ionized impurities density associated to the right-hand QW is set at $N_{d2}=3.5$. Three values of the electric field intensity have been taken to illustrate the deformation of the conduction band bending in the figure 2 a), including negative, zero and positive. However, the optical coefficients contained in the figures 2 b) and 2 c) have been calculated only with the inclusion of negatively oriented static fields. It is straightforward from these results the observance of the increment in the resonant peak amplitudes when the magnitude of the field augments. This is a consequence of the increase in the polarization of the system, induced by the applied field, which reflects in the growth of the transition dipole matrix elements.

136

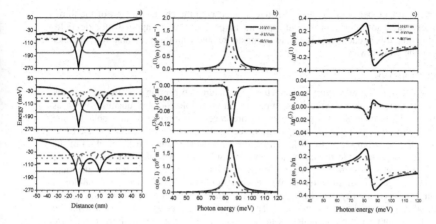

Figure 2. a) Potential energy profile and electronic structure for three different values of the electric field intensity: -10 kV/cm (top), zero (middle), and 10 kV/cm (bottom) and linear and nonlinear optical responses, b) Absorption coefficient, c) Relative change of the refractive index in a GaAs-based double DDQW. In this case, N_{d1}=5.5 and N_{d2}=3.5.

One can also notice that the growth in magnitude of the negatively oriented applied static electric field has the consequence of a blueshift in the position of the resonant peaks. This can be explained with the help of the results appearing in the figure 3, which contains the energy positions of the first four confined energy states in the GaAs double DDQW as functions of the electric field intensity. It can be directly seen that the energy difference for the main transition, $E_1 - E_0$, decreases as the field goes from -10 until 10 kV/cm. In consequence, the observed blueshift is justified by the corresponding increase in E_{10} noticed from the figure 3.

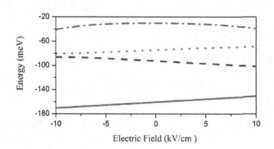

Figure 3: Energy levels as a function of the applied electric field intensity.

The same strange behavior mentioned above, related with the change in sign of the third-order optical absorption contribution, appears in presence of an applied electric field, revealing that this phenomenon seems to be mostly related with the particular system under study in the present work.

CONCLUSIONS

In this work we report the electronic structure, absorption coefficient and the relative refractive index change in asymmetric GaAs double δ-doped quantum wells as functions of the main system parameters in presence and absence of an applied static electric field. We obtain that the energy difference for the main transition, $E_1 - E_0$, decreases as the field goes from -10 until 10 kV/cm. It means that the resonant peak position of the absorption coefficient and the maxima or minima of the relative refractive index change will experience a redshift as functions of the applied electric field as long as the field intensity changes within this interval. We also obtain that this asymmetric double δ-doped system exhibits a non-typical behavior related with the third-order contribution to the optical absorption coefficient considered. This phenomenon appears to be related with the values of the ionized impurity densities in the system and with the strength of the electric field. However, an additional study is required due that it seems that the origin of this unusual property relates mostly with the particular potential configuration of the asymmetric δ-doped double asymmetric quantum well. Finally, we have also calculated the relative refractive index change for the system which, at least for the specific incident light intensity considered, does not exhibits a non-typical behavior.

ACKNOWLEDGMENTS.

The authors would like to acknowledge to the National Science and Technology Council (CONACyT) from Mexico and to the Administrative Department of Science, Technology and Innovation (COLCIENCIAS) from Colombia for the partial financial support trough bilateral collaboration grant B330.500. MEMR is grateful to CONACyT for support under sabbatical Grant 2011-2012 No. 180636.

REFERENCES

[1] C. E. C. Wood. G. Metze, J. Berry, and L. M. Eastman. J. Appl. Phys. **51**, 383 (1980).

[2] L. Ioriatti. Phys. Rev. B. **41**, 8340 (1990).

[3] I. Saidi, L. Bouzaïene, and H. Maaref, J. Appl. Phys. **101**, 094506 (2007).

[4] J. C. Martínez-Orozco, M. E. Mora-Ramos, and C. A. Duque, Phys. Stat. Sol. B **249**, 146 (2012).

[5] J. C. Martínez-Orozco, M. E. Mora-Ramos, and C. A. Duque, J. Lumin. **132**, 449 (2012).

[6] H. Yildirim and M. Tomak. Eur. Phys. J. B **50**, 559 (2006).

Mater. Res. Soc. Symp. Proc. Vol. 1479 © 2012 Materials Research Society
DOI: 10.1557/opl.2012.1611

Hydrostatic pressure effects onto the electronic structure and differential capacitance profile for a metal/δ-doped-GaAs.

A. Puga[1], J. C. Martínez-Orozco[1]

[1] Unidad Académica de Física. Universidad Autónoma de Zacatecas. Calzada Solidaridad esquina con Paseo a La Bufa S/N. C.P. 98060, Zacatecas, Zacatecas, México.

ABSTRACT

A metal-semiconductor contact with a n-type δ-doped quantum well of impurities (metal/δ-doped GaAs) was studied numerically to extract electronic properties such as energy levels and the corresponding wave functions of each level as well as the differential capacitance for the structure. In this work we reported these properties as a function of the hydrostatic pressure (P). We used the effective mass approximation for the calculation of the electronic structure and consider the hydrostatic pressure effects on the basic semiconductor parameters as is the effective mass for the conduction electrons and the static dielectric constant, finally we also take into account an experimental expression that dependency of the Schottky barrier height as a function of this external factor, at least for values between 0 and 6 kar. We showed that the linear behavior for C^{-2}, obtained by pervious works, is switched over a parabolic-like curve due to the δ-doped two-dimensional impurities density.

Keywords: electronic structure, nanostructure, III-V.

INTRODUCTION

Metal-semiconductor contacts are one of the most elementary components (building blocks) of the semiconductor devices; one widely studied example is the metal/GaAs contact due to its great impact on electronics technology. Recently, scientists have added to the semiconductor devices a single or multiple δ-doped quantum well (DDQW) of impurities, that consist in the incorporation of a relatively high number of impurities in a single semiconductor layer, in order to improve the device performance due to the huge amount of charge carriers that it provides to the devices. In this work we numerically obtained the electronic structure of metal/ δ-doped GaAs system as well as the differential capacitance profile considering hydrostatic pressure effects. The numerical simulation is carried out within the effective mass approximation by using a simple model for the potential of the conduction band of the system. The influence that the hydrostatic pressure has in our potential is considered through the system parameters dependency on the hydrostatic pressure (P), for example the effective mass ($m^*(P)$), the static dielectric constant $\epsilon(P)$ and the Schottky Barrier Height $V_C(P)$.

In this work we reported the hydrostatic pressure effects on the electronic structure as well as in the differential capacitance per unit of area. We showed results for a pressure range of 0-6 kbar and different n-type δ-doped quantum well impurities density (N_{2d}). It is reported the dependency with the hydrostatic pressure of energy levels, potential profile and differential capacitance curves.

THEORY

The way we treat this problem is through a simple model for the potential profile, which consider the Schottky barrier potential related to the Metal-Semiconductor contact in the gate terminal and also the band bending due to the incorporation of the δ-doped sheet of impurities. The analytical formula for the potential profile is given by [3]:

$$V(z) = \frac{4\pi}{\varepsilon_s} N_d (z - l)^2 \theta(z - l_d) + V_n(z)\theta(z - l_d).$$

The first term in this expression for this potential profile is the parabolic Schottky barrier that is found by working in the depletion region approximation, the second term represents the potential due to the δ-doped sheet of impurities that was originally obtained by Ioriatti [4] within the Thomas-Fermi approximation being θ the heavyside step function and l_d the depletion region width. In this work we also consider an exchange and correlation contribution to the Hartree potential proposed by Gaggero-Sager [5]. So the total $V_n(z)$ is given by:

$$V_n(z) = -\frac{\alpha_n^2}{(\alpha_n|z| + z_{0n})^4} - c\left[1 + \frac{a(\alpha_n|z| + z_{0n})^2}{\alpha_n}\ln\left(1 + \frac{b\alpha_n}{(\alpha_n|z| + z_{0n})^2}\right)\right]\frac{\alpha_n}{(\alpha_n|z| + z_{0n})^2}.$$

The main involved parameters are: $\alpha_n = 2/15\pi$, $z_{0n} = (\alpha_n^3/\pi N_{2d})^{1/5}$, $a = 0.7734\alpha_n/21$, $b = 21\alpha_n^{-1}$, $c = 2/\pi$, where N_{2d} represents the two-dimensional impurities density for the n-type δ-doped quantum well, the other potential model parameter are in [3]. In figure 1 we present the computed energy electronic structure for the system. In this particular potential profile: $N_{2d} = 7.5 \times 10^{12}$ cm^{-2}, the distance from the Metal-Semiconductor contact to the δ-doped sheet of impurities is of 30 nm, the contact potential in this case is of 500 meV.

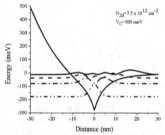

Figure 1: Potential profile and probability density for the n-type δ-doped system in Gallium Arsenide (GaAs). In this case: $N_{2d} = 7.5 \times 10^{12}$ cm^{-2}, the distance from the Metal-Semiconductor contact to the δ-doped sheet of impurities is of 30 nm and the contact potential is of 500 meV.

The consideration of the hydrostatic pressure effects onto the electronic structure in the δ-doped well is done through the dependency of the basic physics parameters in the effective mass approximations [6]: The hydrostatic pressure (P) dependency of the effective mass:

$$\frac{m(P)}{m_0} = \left[1 + \frac{15020}{1519 + 10.7P} + \frac{7510}{1519 + 10.7P + 341}\right]^{-1}$$

where m_0 is the free electron mass.

The static dielectric constant as a function of P is

$$\epsilon(P) = 12.82 \, e^{\left(-1.67 \times 10^{-3} \, kbar^{-1} P\right)}$$

Finally, as we already mention that in the introduction, there exists an experimental work from Cancaya et al. [1] where they reported the effect that the hydrostatic pressure effect has on the Schottky barrier height, this is an analytical expression obtained from experimental measurements given explicitly by:

$$Vc(P) = Vc(P = 0) + \gamma P + \kappa P^2 + \rho P^3,$$

where $\gamma = 11.21 \, meV/kbar$, $\kappa = -0.345 \, meV/kbar^2$, $\rho = 0.25 \, meV/kbar^3$. This is the total simple model for the potential profile that permits us to calculate not only the energy level structure by solving a Schrodinger-like equation, we also are able to compute some other important properties as optical properties as reported in [6, 7]. In this work we are interested not only in the energy level structure, we are investigating the hydrostatic pressure effect on the differential capacitance profile, in fact, we already perform a previous work in this line [3], but in this work we investigate the effect of hydrostatic pressure on the differential capacitance profile. The expression used to compute the differential capacitance per unit of area is given by:

$$C^{-2}(P) = \frac{2V_c(P)}{\epsilon(P)eN_d}$$

Where V_C is the contact potential in the Metal-Semiconductor Contact and N_d is the background impurities density an in this work is of $N_d = 1.0 \times 10^{18} \, cm^{-3}$.

RESULTS AND DISCUSSION
We have studied the influence that the hydrostatic pressure has in the electronic level structure of a Metal/δ-doped-GaAs system as well as in the differential capacitance profile. We are interested in the sensitivity –or effect- on the energy levels with the applied hydrostatic pressure, for this to be theoretically investigated we chose a particular set of parameters, those reported in figure 1, and we carry on some numerical simulations.

First, in figure 2 a) we report the energy levels as a function of the hydrostatic pressure for different values of the contact potential (300, 400 and 500 meV). In this case we found that for values of the contact potential barrier below 300 meV, the hydrostatic pressure does not affect that ground state (E_0) of the system, but as we increase the value of V_c the hydrostatic pressure effects becomes appreciable, at least for the ground state for P=5 kbar, we finally stress the fact that for the third exited state (E_3) experiences an almost constant energy shift due to hydrostatic pressure effects. Then, in figure 2 b) we calculate the energy level dependency on the contact potential value for three different values on the of hydrostatic pressure: P=0 (Continuous line), P=4 kbar (dashed line) and P=8 kbar (point-dashed line).

a)   b)

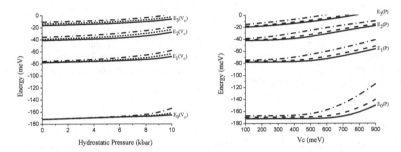

Figure 2: a) Energy levels as a function of the hydrostatic pressure for different values of the contact potential; $V_C = 300$ meV (Continuous line), $V_C = 400$ meV (dashed line) and $V_C = 500$meV (point-dashed line). In b) we report the energy levels as a function of the contact potential for different values of hydrostatic pressure: $P = 0$ kbar (Continuous line), $P = 4$ kbar (dashed line) and $P = 8$ kbar (point-dashed line).

Finally we perform the differential capacitance as a function of the contact potential simulation with the methodology reported in [3] that fundamentally consist in calculate the infinitesimal variation of the depletion region with an infinitesimal variation of the applied contact potential for the system. In fact we report C^{-2} vs V_C - instead of V_C vs V_C, because for the Metal-Semiconductor contact this curve is a line whose slope is proportional to the background impurities density, but when we introduce the δ-doped system this is switched over a parabolic-like curve, that also depends on the value of P as can be seen on Figure 3.

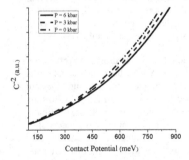

Figure 3: Differential capacitance profile C^{-2} - in arbitrary units - as a function of the contact potential V_C for three different values of hydrostatic pressure P. The point-dashed line corresponds to a non-applied hydrostatic pressure; the dashed and solid curves correspond to 3 and 6 kbar, respectively.

CONCLUSIONS

In summary, we successfully derived the electronic structure for the metal/δ-doped-GaAs system under hydrostatic pressure. We found the relationship between the electronic level structure and the hydrostatic pressure under the variation of several parameters such as the contact potential. With respect to the differential capacitance profile, as proposed in [3], the presence of the two-dimensional electronic gas (TDEG) due to the δ-doped sheet of impurities produces a diminishing of the depletion region width as a function of the contact potential that give rise to a non-linear behavior of the differential capacitance (C^{-2} vs V_C) per unit of area of the device, as expected for a Metal/GaAs system $i.e$ without δ-doped system. Here we also showed that this parabolic-like curve also depend on the applied hydrostatic pressure.

ACKNOWLEDGMENTS

The authors would like to acknowledge to the Integral Strengthening Institutional Program (PIFI) from Mexico for partial financial support for this work.

REFERENCES

[1] G. Çankaya, N. Uçar, E. Ayyildiz, H. Efeoğlu, A. Türüt, S. Tüzemen, and Y. K. Yoğurtçu. Phys. Rev. B. **60**, 15 944 (1999).

[2] M. E. Mora-Ramos and C. A. Duque, Brazilian Journal of Physics. **36**, 866 (2006).

[3] J.C. Martínez-Orozco, L.M. Gaggero-Sager, Stoyan J. Vlaev, Solid-State Electronics **48**, (2004) 2277.

[4] L. Ioriatti. Physical Review B. **41**, 8340 (1990).

[5] L. M. Gaggero-Sager. Modelling and Simulation in Materials Science and Engineering. **9**, 1 (2001).

[6] J. C. Martínez-Orozco, M. E. Mora-Ramos, C.A. Duque, Journal of Luminescence. **132**, 449 (2012).

[7] J.C. Martínez-Orozco, M. E. Mora-Ramos, C.A. Duque. physica status solidi B. **249**, 146 (2012).

AUTHOR INDEX

SUBJECT INDEX

acoustic, 21
adhesive, 89
adsorption, 51
Ag, 51, 57
alloys, 1, 9, 89
Au, 1

biological, 27, 107
biomedical, 63

catalytic, 77
Chemical synthesis, 45, 69
chemical vapor deposition (CVD),
   111
composites, 51, 57, 101
crystal growth, 95
Cu, 15

elastic properties, 89
Electrical properties, 117
electronic structure, 15, 107, 117,
   125, 133, 139

ferroelectric, 33
fiber, 33

hydrogenation, 39

III-V, 125, 133, 139

morpholoy, 83

nano-structure, 21
nanoscale, 21, 63
nanostructures, 1, 9, 15, 33, 39, 57,
   77, 83, 95, 101, 117, 139
Ni, 39
nitride, 107

optical properties, 125, 133
optical, 27
Organic, 45, 69

Pd, 1
polymerization, 51

Raman Spectroscopy, 107

scanning electron microscopy
   (SEM), 111
sensor, 27, 83, 95
simulation, 1, 111
sol-gel, 9, 63
strength, 89

thermal conductivity, 101

W, 77

ZnO, 89

Printed in the United States
by Baker & Taylor Publisher Services